MATHEMATICAL
DIVERSIONS

MATHEMATICAL
DIVERSIONS

J. A. H. HUNTER

and

JOSEPH S. MADACHY

DOVER PUBLICATIONS, INC.
NEW YORK

Published in Canada by General Publishing Company, Ltd., 30 Lesmill Road, Don Mills, Toronto,
Ontario.
Published in the United Kingdom by Constable
and Company, Ltd., 10 Orange Street, London WC 2.

This Dover edition, first published in 1975, is an
unabridged and revised republication of the work
originally published by D. Van Nostrand Company,
Inc., Princeton, New Jersey, in 1963, with a new
preface by the authors.

International Standard Book Number: 0-486-23110-0
Library of Congress Catalog Card Number: 74-83619

Manufactured in the United States of America
Dover Publications, Inc.
180 Varick Street
New York, N. Y. 10014

PREFACE TO THE DOVER EDITION

In this new Dover edition we have incorporated recent discoveries such as details of the new greatest known prime number. Apart from these, and the correction of certain errors, there have been no changes from the original text.

August 1974   THE AUTHORS

PREFACE TO THE FIRST EDITION

"Something old, and something new"! This seem to sum up what we have tried to achieve in this book. And with emphasis on the fun that the true lover of recreational mathematics finds in doing, rather than in reading about the doing!

Beyond that there is little need for introductory remarks so far as the content is concerned.

For much of the information on polyominoes we are indebted to S. W. Golomb's pioneering work. The valuable suggestions of a few kind people have been acknowledged individually in the text. For various other ideas, we would like to express our thanks in particular to the following:

Jean H. Anderson	Sinclair Grant
Stephen Barr	Guy Guillotte
A. G. Bradbury	R. B. McDunphy
Alan L. Brown	Derrick Murdoch
Spencer Earnshaw	John Waldorf

January, 1963   THE AUTHORS

CONTENTS

Chapter 1

FRIENDLY NUMBERS AND OTHERS

Any numerologist will tell you that numbers, **all** numbers, have personalities. However, you don't have to believe in numerology to believe in the personalities of **some** numbers. For example, how about the number 13? One immediately conjures up some thought connected with bad luck—an obnoxious personality has attached itself to this number 13. Numerically, 13 is just one of the smaller prime numbers (though it does have some special interest when we discuss Mersenne primes later).

Professional and amateur gamblers will tell you that the number 7 is schizophrenic—good luck at the right time, bad luck at the wrong time. The number 3 has had quite a reputation for being somewhat mystical—the Trinity, the 3 hours Christ hung on the Cross, the 3 days He lay in the tomb; Pythagoras called it the "perfect number" expressive of the "beginning, middle, and end" making it the symbol of Deity; Ancients considered the world ruled by 3 gods—Jupiter (in Heaven), Neptune (in the sea) and Pluto (in Hell); then there are the 3 Fates, the 3 Graces, the Muses which were 3 times 3; and Man is considered as body, soul and spirit. Similarly the number 7 has had a reputation for being mystical— the 7 days of creation, the 7 deadly sins, the ancient Hebrew "7 names of God." Here are two numbers, out of many, with definite religious associations.

From time immemorial the odd numbers have been endowed with male, the even numbers with female attributes, and *vice versa*, according to the culture concerned.

But these personalities associated with numbers are man-made and result from human experiences. There are numbers that have personalities inherent in the numbers themselves.

Some numbers are uninteresting in themselves[1] (except to a con-

---

[1] Notwithstanding the well-known argument which demonstrates that there are no uninteresting numbers! To wit: let us assume that there are two classes of numbers—interesting and uninteresting. Certainly all numbers could be put into

1

firmed numerologist) and acquire some status only when related to other numbers. The number 220 doesn't appear to have any special significance. But add up all its integer divisors, except the number itself:

$$1 + 2 + 4 + 5 + 10 + 11 + 20 + 22 + 44 + 55 + 110 = 284$$

Now do the same to the number 284:

$$1 + 2 + 4 + 71 + 142 = 220$$

We seem to have something here! 220 and 284 are intimately related since each is the sum of the divisors of the other. As a matter of fact, such number pairs are called *amicable* numbers, which means "friendly" numbers. Not too many of these number pairs are known—less than 1200—and the great Euler, in 1750, discovered 59 of them himself. Some other amicable number pairs are 1184 and 1210, 2620 and 2924, 5020 and 5564, 17,296 and 18,416, 9,363,584 and 9,437,056. The smaller pair, 220 and 284, has been known from antiquity and so much significance was attached to it that the possessor of one (in the form of talismans, or numerological significance) was assured of close friendship with the possessor of the other number of the pair. Undoubtedly, some marriages have been made on the basis of amicable numbers (considering the bases of some modern marriages, amicable number pairing may have had its points!).

The most studied class of numbers with definite inherent personalities is the class of prime numbers. Primes are integers which have no integer divisors, except themselves and 1: 2, 3, 5, 7, ... 229, ... 5693, ... 199,999 and so on out to infinity. Prime numbers have rather defiant personalities and the larger the prime the more apparent is this defiance. Who cares that such a small number as 23 is prime? But how about 10,000,019 which has absolutely no integer divisors? Even this number is puny and insignificant compared to some of the LARGE primes which will be mentioned later. Then there are some primes that look incredible:

---

one or the other of these classes. Now, in the class of uninteresting numbers there is a largest and a smallest number. Obviously, this makes them interesting! If we transfer these to the class of interesting numbers we will again have a largest and smallest uninteresting number, again making them interesting. Eventually we are left with one or two uninteresting numbers. But since they are the *only* uninteresting numbers, they are interesting for that reason! *Reductio ad absurdum.*

$$1,111,111,111,111,111,111$$
$$11,111,111,111,111,111,111,111$$
$$909,090,909,090,909,090,909,090,909,091$$
$$9,090,909,090,909,090,909,090,909,090,909,091$$

There are many and various methods available to test a given number to see if it is prime. For the very large primes special methods have been devised which, even then, must utilize electronic computers. Basically, however, the method involves dividing the given number by all the primes less than the square root of the given number. If we were to test 233 we would divide by all the primes less than $\sqrt{233}$, i.e., those primes less than 15. Dividing by 2, 3, 5, 7, 11 and 13 we find a remainder each time, so 233 is a prime. To test 5659 by this method we would have to divide by all the primes less than $\sqrt{5659}$, i.e. those less than 75. This involves 21 primes—though various tricks can obviate some of the divisions. But all the tricks in the trade won't make an easy task of determining the primality of a number like 8,083,457 which would involve division by the 412 primes less than $\sqrt{8,083,457}$. And who could ever use this method to check out the primality of those four incredible prime numbers given above?

Prime numbers have been of major mathematical importance since Euclid, about 2300 years ago, showed that there was no limit to the number of primes. He argued roughly as follows: Suppose there were a largest prime P. The product of all the primes, plus 1, is either prime or non-prime (composite), i.e.,

$$(p_1 p_2 p_3 \ldots P) + 1 = \text{a prime or a composite number}$$

This number is not divisible by any of the primes smaller than P, or by P itself, since the remainder of 1 would result. Therefore, the number is either a prime, or it is a composite number with a prime factor larger than P. In either case, the existence of a prime greater than P is demanded. Therefore, there is no largest prime P—only a largest *known* prime.

Various tables of primes have been compiled. One of the most extensive of these is an 8-volume tabulation of the primes to 100,-330,201, which suffers from many errors (and its second volume has been lost). The original manuscript is preserved in Vienna and is still an extremely valuable document in spite of its inadequacies. Kulik, who spent much of his life on this table, listed over 5,761,456 primes. A nearly perfect table was published by Derrick N. Lehmer in 1914, and this includes all the primes up to and including 10,006,721—a list of 665,000 numbers including 1 (not usually considered a prime). Recently, microcard copies of the first 6,000,000 primes have been published. They were calculated by C. L. Baker and F. J. Gruenberger of the RAND Corporation on an IBM704 computer. This extensive table covers the primes up to 104,395,289.

Early Chinese manuscripts show that masculine qualities were ascribed to primes, although all other odd numbers were considered feminine. Even in those ancient days prime numbers were recognized as being very special, and now we have seen that modern mathematicians still feel that way about them.

In addition to the nearly 6,000,000 primes listed in the tables mentioned above, isolated larger primes of many types are known and most have some history attached to them—considering the tremendous task involved in establishing the primality of some of the larger primes. There is a group of prime numbers known as Robinson numbers given by the formula $R(k,n) = 2^n k + 1$ which yields primes for certain values of k and n. For $k = 5$ and $n = 1947$, the prime number formed contains 586 digits and is the largest known Robinson prime, though by no means the largest known prime.

Another formula which gives a few primes is one devised by Fermat, $2^{2^n} + 1$. Fermat firmly believed that this would yield primes for all values of n, but he was very wrong. Only five primes have been discovered which conform to this formula: 3, 5, 17, 257 and 65,537 from values of $n = 0$, 1, 2, 3 and 4 respectively. The very next value, $n = 5$, yields 4,294,967,297 which has the two prime factors 641 and 6,700,417. The compositeness of some Fermat numbers has been established, but no further primes have been discovered. The largest Fermat number tested is $2^{2^{1945}} + 1$ which contains about $10^{10^{584}}$ digits! Indeed, it is one of the largest numbers tested for primality. The complete factorization of this number is not known, but one of its prime factors is the large 586-digit Robinson prime $R(5, 1947)$. Since this large Fermat number, symbolized as F_{1945}, has been shown to be non-prime, it is doubtful that any further interest will be shown in establishing any of the other prime factors it may contain. There are limits to the attraction such numbers can have for even the most ardent number theorist.

The Fermat numbers have a close relationship with certain geo-metrical constructions, viz., the construction of regular polygons of n sides, where n is an odd prime. Although there is no difficulty in constructing equilateral triangles or regular pentagons, there are no methods available (using only a straightedge and compass) for constructing regular heptagons, or 11-gons, and others. A construc-tion for a 17-gon is known and Carl Friedrich Gauss, in 1798 at about the age of 19, made the startling discovery that such con-structions are possible if and only if n, the number of sides of the regular polygon, is a Fermat prime. In other words, Euclidean construction of a regular polygon with a prime number of sides is only possible if the number of sides is 3, 5, 17, 257 or 65,537.

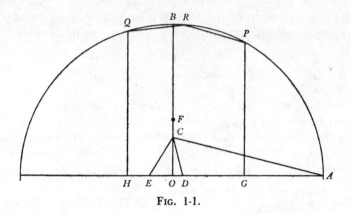

Fig. 1-1.

Equilateral triangles and regular pentagons are easy to draw. A regular 17-gon can be constructed as in Figure 1-1: [2]

Construct a semicircle with center at O and radius OA and draw a perpendicular OB. From A draw AC where OC is $\frac{1}{4}OB$. Con-struct $\angle OCD = \frac{1}{4} \angle OCA$ and $\angle ECD = 45°$. With EA as a diam-eter draw a semicircle which cuts OB in F. With D as center and DF as radius draw another semicircle and draw perpendiculars G and H where this semicircle cuts OA. P and Q, where the perpen-diculars cut the large semicircle, form the extremities of an arc equal to 2/17 of the circumference. Bisection of this arc gives us R. PR or RQ is one side of the 17-gon.

[2] This is adapted from H. S. M. Coxeter, *Introduction to Geometry*, John Wiley & Sons, Inc., 1961, page 27.

Constructions for a 257-gon and a 65,537-gon are known, but would be rather too lengthy to detail here. It is interesting to note that it took O. Hermes ten years to construct the regular 65,537-gon and that his manuscript, filling a large box, is at the University of Göttingen. Mathematics certainly brings out the hidden resources of some individuals—though the usefulness of such endeavor is, at times, questionable.

Now, getting back to prime numbers. Let us examine some of the various other formulae which have been devised to generate limited series of primes. The most well-known is Euler's polynomial $x^2 - x + 41$ which gives 40 different primes for $x = 1, 2, 3, \ldots 40$. The very simple $2x^2 + 29$, found by Legendre in 1798, generates 29 primes for $x = 0, 1, 2, \ldots 28$. No *polynomial* can generate primes exclusively, but here is a formula devised by Malcolm H. Tallman which, when properly used, generates primes exclusively:

$$N = \frac{P_n}{a_i a_k \ldots a_n} \pm a_i a_k \ldots a_n$$

where (1) P_n is the product of the first n primes, and $a_i, a_k, \ldots a_n$ are any of the first n primes, all different, and/or unity and (2) N is less than the square of the $(n + 1)$th prime. Then every such N is a prime.[3]

An example will make the use of this formula clear:

$$N = \frac{2 \cdot 3 \cdot 5 \cdot 7 \cdot 11 \cdot 13 \cdot 17 \cdot 19 \cdot 23}{2 \cdot 3 \cdot 11 \cdot 13 \cdot 17} - (2 \cdot 3 \cdot 11 \cdot 13 \cdot 17) = 709, \text{ a prime}$$

because $709 < 29^2$.

The proof that no polynomial can generate primes exclusively might be included at this point since it is not often seen by the general reader.

If a rational algebraical formula can represent only prime numbers, we can generalize such a formula as:

$$a + bx + cx^2 + dx^3 + \ldots.$$

Then, if P be a prime number derived from this formula when $x = m$, we have:

$$P = a + bm + cm^2 + dm^3 + \ldots.$$

[3] The proof of Tallman's formula was published in *Recreational Mathematics Magazine*, No. 4, August 1961, page 64.

Also, if we set $x = m + nP$, we must obtain another prime number, say Q, with value:

$$Q = a + b(m + nP) + c(m + nP)^2 + d(m + nP)^3 + \ldots$$

Now,
$$b(m + nP) = bm + \text{(terms containing } P)$$
$$c(m + nP)^2 = cm^2 + \text{(terms containing } P)$$
$$d(m + nP)^3 = dm^3 + \text{(terms containing } P)$$

etc.

So,
$$Q = a + bm + cm^2 + dm^3 + \ldots + \text{(terms containing } P)$$
$$= P + \text{(terms containing } P)$$

The expression (terms containing P) must be a multiple of P, hence Q must be a multiple of P and so cannot be prime. This shows that a rational algebraical formula cannot only represent primes.

Prime pairs are those primes which differ by 2, e.g., 3 and 5, 11 and 13, 19,469 and 19,471. There are not many polynomials or formulae which yield prime pairs, but here is one discovered by A. T. Gazsi:

$$N = 12{,}150 - 1710A + 60A^2$$

where A takes successive values 1, 2, 3, ... 20 and $N + 1$ and $N - 1$ are the prime pairs. 18 pairs of positive prime pairs are generated and one pair of negative primes (-29 and -31).

There are still many unsolved problems and unproven conjectures in connection with prime numbers.

Bertrand's Postulate states that there is at least one prime between N and $2N$, where N is any integer >1. This has been proven, but no formula has been found for the exact number of primes in that interval: there is also a proof that, very roughly speaking, there are as many primes between N and $2N$ as there are between zero and N. In any particular case the exact number of primes can be found only by actual trial: for example, between 12 and 24 there are four primes (there are five primes between zero and 12). Applying Bertrand's Postulate we know that there must be at least three primes containing exactly 100 digits—and three are now known: $35 \cdot 2^{327} + 1$, $63 \cdot 2^{326} + 1$, and $81 \cdot 2^{324} + 1$.[4]

---

[4] Using Tchebycheff's formula $\int_{2}^{x} (\log x)^{-1} dx$ for the approximate number of primes less than x, we find that there are very roughly 3.9×10^{97} primes between 10^{99} and 10^{100}.

Many arithmetical series of primes are known: for example, 11, 17, 23, 29 which has a common difference of 6. A longer series of 10 terms with a common difference of 210 starts with 199. The longest known arithmetical sequence of primes, discovered by S. C. Root in 1969, contains 16 terms—starting with 2,236,133,941 with a common difference of 223,092,870. Mathematicians feel there must be longer prime arithmetical series than these, but the methods available for finding these series are hardly better than trial and error or hunting in the tables of primes. A question that remains unanswered is: Is there a series of primes in arithmetical order of arbitrary length? Is there, for example, an arithmetical series of primes of 50 terms?

Goldbach's Theorem states that every even number is the sum of, at most, two primes. Though actual trial has never yielded a refutation of this, the closest we have been able to come to a proof is Vinogradoff's proof that there is an integer N such that any integer $n > N$ can be represented as the sum of not more than three primes if n be odd, four primes if n be even. However, we have no idea as to what the limit of that N may be for either case.

We gave Euclid's proof that there are infinitely many primes, but it has not yet been determined whether or not there are infinitely many prime pairs. It seems probable that there are, but this remains a mere conjecture at present.

The most famous class of prime numbers are Mersenne primes, named after Marin Mersenne who announced in 1644 that he had discovered some new Perfect numbers (more about Perfect numbers later). In so doing he stated that the form $2^p - 1$ is prime for $p = 2, 3, 5, 7, 13, 17, 19, 31, 67, 127$ and 257. He was wrong on five counts: 67 and 257 do not yield primes and he missed 61, 89 and 107 which do yield primes. In recognition of his efforts, however, his name has been attached to primes of form $2^p - 1$ symbolized by M_p. Since Mersenne made his announcement, several other greater M_p have been found. The complete list to date is given by the following values of p: 2, 3, 5, 7, 13, 17, 19, 31, 61, 89, 107, 127, 521, 607, 1279, 2203, 2281, 3217, 4253, 4423, 9689, 9941, 11213, 19937, 21701, 23209, 44497. The last value gives a prime $2^{44497} - 1$, with 13395 digits. Even with a computer one could not test such a huge number for primality by the previously mentioned method of laboriously dividing successively by all primes less than

$\sqrt{2^{19937}-1}$. A special method was devised by Lucas in 1876 to test the primality of Mersenne numbers. Consider the series 4, 14, 194, 37,634, ... in which each term is 2 less than the square of the preceding term (i.e. $S_n = S_{n-1}^2 - 2$). If the $(p-1)$th term of this series is divisible by $2^p - 1$ without a remainder, then M_p is a prime. When you consider that the fifth term of this series is 1,416,317,954 you can readily imagine the size of the 19937th term! But mathematicians would not spurn a convenient test even though it appeared to be more cumbersome than the problem it was devised to overcome. As soon as a term of the Lucas' series gets larger than the M_p being tested, the term is divided by M_p and the remainder, if any, is used to continue the series. For example, suppose we wanted to test the primality of $2^7 - 1 = 127$, using Lucas' series.

> The first term is 4
> The second term is $4^2 - 2 = 14$
> The third term is $14^2 - 2 = 194$

Since 194 is greater than 127, we divide by 127 and get a remainder of 67. We then continue:

> The fourth term is $67^2 - 2 = 4487$; divide by 127, get a remainder of 42,
> The fifth term is $42^2 - 2 = 1762$; divide by 127, get a remainder of 111,
> The sixth term is $111^2 - 2 = 12321$; divide by 127, get *no* remainder. Therefore 127 is prime.

Even with Lucas' series no one would tackle the larger Mersenne numbers without the assistance of an electronic computer. The largest known prime is a Mersenne prime: M_{44497} containing 13395 digits—a truly defiant number it is!

What makes all this even more amazing is the fact that for every Mersenne prime there corresponds a Perfect number—and there is no known Perfect number which does not correspond to a Mersenne prime.

Perfect numbers, in contrast to prime numbers which have no integer divisors, **do** have integer divisors—but only enough to ensure that their sum is equal to the number itself. The divisor 1 is included, but the number itself is excluded. For example:

$$6 = 1 + 2 + 3$$
$$28 = 1 + 2 + 4 + 7 + 14$$

6 and 28 are two Perfect numbers. Others are 496, 8128 and 33,550,336. A total of 24 Perfect numbers are known and all are even numbers.[5] The first four Perfect numbers have been known for 2000 years but in 1460 some anonymous person recorded the fifth Perfect number. No one has found any odd Perfect numbers and there is doubt that any exist. Euler proved, in 1750, that all even Perfect numbers are of the form known to Euclid, viz., 2^{p-1} $(2^p - 1)$ where p and $2^p - 1$ are prime ($2^p - 1$ is our old friend, the Mersenne prime). For each of the Mersenne primes defined on page 8, there corresponds a Perfect number given by the formula above. The 27th Perfect number has 26790 digits!

One might wonder why mathematicians apparently waste their time on prime and Perfect numbers. However, the study of these has led incidentally to great advances in number theory. One of the unsolved problems in number theory is whether or not there is an infinity of Mersenne primes and, therefore, an infinity of Perfect numbers. Concerning the existence of odd Perfect numbers, we only do know that if odd Perfect numbers exist, they must be of the form $12m + 1$ or $36m + 9$ where m is prime. Oystein Ore writes "It has been shown by H. J. Kanold that there are no odd Perfect numbers smaller than 1.4×10^{14}. One of my students, J. B. Muskat, has informed me that he has been able to raise this lower limit to 10^{18}."

Proving that there is *not* an infinity of Mersenne primes would not settle the question of an infinity of Perfect numbers until the problem of odd Perfect numbers is solved. One day, perhaps, someone will discover the first odd Perfect number, or evolve proof of its non-existence. One day there may be another step forward regarding Mersenne primes. With infinity as a goal, it is not likely that enthusiasm for finding the next Mersenne prime will diminish. And the concomitant results can be only the advancement of number theory and mathematics in general.

Various interesting facts are known about Perfect numbers:

---

[5] There is a class of *Multiperfect* numbers in which the sums of the divisors of a number is a *multiple* of the number. The sum of the divisors of 120 (1, 2, 3, 4, 5, 6, 8, 10, 12, 15, 20, 24, 30, 40, 60) equals 240. (Mersenne first pointed out the existence of this multiperfect number in 1631.) Other numbers, considerably greater in size, are known which are multiperfect of the order 3, 4 or even 7 in which the sums of the divisors are 3, 4 or 7 times the number itself.

All known Perfect numbers, except 6, have digital roots of 1, i.e., the ultimate sum of their digits equal 1.

e.g., 496: $4 + 9 + 6 = 19$ $1 + 9 = 10$ $1 + 0 = 1$

Every known Perfect number, except 6, is the sum of consecutive odd cubes, beginning with 1.

$$28 = 1^3 + 3^3$$
$$496 = 1^3 + 3^3 + 5^3 + 7^3$$
$$8128 = 1^3 + 3^3 + 5^3 + 7^3 + 9^3 + 11^3 + 13^3 + 15^3$$

and so on where the number of consecutive odd cubes is equal to $\sqrt{2^{p-1}}$.

Perfect numbers are the sums of the successive powers of 2 from 2^{p-1} to 2^{2p-2}:

$$6 = 2^1 + 2^2$$
$$28 = 2^2 + 2^3 + 2^4$$
$$496 = 2^4 + 2^5 + 2^6 + 2^7 + 2^8$$

There are many other special properties associated with Perfect numbers, but those mentioned here will have shown that the long-standing interest in them throughout the ages has not been without good cause. And, having seen what can be made of just a few numbers in general, it becomes easier to understand some of the superstitions that were built up around them. Perhaps the numerologists are right after all—numbers do have personalities!

Chapter 2

FROM PARADOX TO PARASTICHY

The checkerboard paradox is well known, but it will serve as a lead to some interesting and quite surprising conclusions.

Here we have the paradox in its familiar form, using a skeleton checkerboard 8 units square. Note that the squares are not colored for this. When the board is cut up into four pieces as shown (Figure 2-1), and the pieces then re-arranged to form the 5 by 13 rectangle (Figure 2-2), we seem to gain one unit. The truth, of course, is that the final long "diagonal" of that rectangle is not in fact a true line: it is a long and narrow space, a sort of very flattened diamond shape, with an area of exactly one unit.

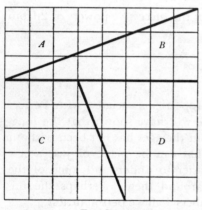

FIG. 2-1.

Now we take a cruder version of the same paradox, using a board only 5 units square (Figure 2-3).

Here we seem to have lost a unit in forming the 3 by 8 rectangle (Figure 2-4). In fact, however, there is an overlap along the line

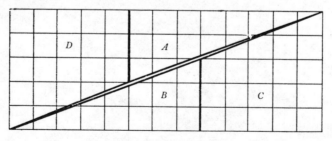

FIG. 2-2.

12

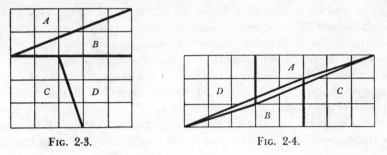

FIG. 2-3.   FIG. 2-4.

of the diagonal in this rectangle and that overlap accounts exactly for the missing unit.

In one case we apparently gained one unit, in the other we apparently lost one unit. But notice the similarity between the dissection and reconstruction procedures in the two cases (Figures 2-1 to 2-4). And notice the numerical breakdowns for the two cases:

$$2 + 3 = 5 \qquad 3(3 + 5) = 24 = 5^2 - 1$$
$$3 + 5 = 8 \qquad 5(5 + 8) = 65 = 8^2 + 1$$

If you try to demonstrate this paradox, using a square "checkerboard" with more than 8 units to the side, you will find that the next larger square must contain 169 units, i.e. 13 by 13. And, although its construction would be a tedious chore, the next one after that would have to contain 441 units, i.e. 21 by 21. For these we would have the numerical breakdowns:

$$5 + 8 = 13 \qquad 8(8 + 13) = 168 = 13^2 - 1$$
$$8 + 13 = 21 \qquad 13(13 + 21) = 442 = 21^2 + 1$$

By analogy, then, you might expect the essential condition for this paradox to be that you use a square with one or other of the following numbers of units to the side:

$$5, 8, 13, 21, 34, 55, \text{etc.}$$

This is indeed the case, and these are successive numbers in the famous Fibonacci series

$$1, 1, 2, 3, 5, 8, 13, 21, 34, 55, 89, \text{etc.}$$

in which each number (after the second) is the sum of the preceding two numbers. Expressed mathematically,

$$f_1 = f_2 = 1, \qquad f_n = f_{n-1} + f_{n-2} \qquad (n > 2)$$

The series is named from Leonardo of Pisa, nicknamed Fibonacci. In the year 1202, so the story goes, he evolved the sequence of numbers in solving a practical problem connected with the breeding of rabbits.   It does seem likely that he noticed the recurrence relation between successive numbers in the series, but no record of this has come to light.   In fact, the first written mention of this relation appeared only four hundred years later.   And more than two hundred more years passed after that before the first known appearance of the paradox itself in a mathematical journal published in Leipzig in 1868.

The paradox depends on the property of the Fibonacci series,

$$f_{n-2}f_n = f_{n-1}^2 - (-1)^n$$

where f_{n-2}, f_{n-1}, f_n are three successive numbers in the series as already defined.

We have seen the application of this identity in the particular cases of $n = 6$, 7, 8, and 9.   The proof of its validity for all cases (subject to $n > 2$) is given in the Appendix.

There is far more than that paradox to be considered in connection with the Fibonacci series.   So we'll now look at the ratios of the first few successive pairs of numbers in the series, as listed here to 4 four decimal places:

$$\frac{1}{1} = 1.0000 \qquad \frac{2}{1} = 2.0000$$
$$\frac{3}{2} = 1.5000 \qquad \frac{5}{3} = 1.6667$$
$$\frac{8}{5} = 1.6000 \qquad \frac{13}{8} = 1.6250$$
$$\frac{21}{13} = 1.6154 \qquad \frac{34}{21} = 1.6190$$
$$\frac{55}{34} = 1.6176 \qquad \frac{89}{55} = 1.6182$$
$$\frac{144}{89} = 1.6180 \qquad \frac{233}{144} = 1.6181$$

As we work up through the series, this ratio obviously approaches a limiting value which lies between 1.6180 and 1.6181.   It can be shown that this limiting value is indeed $\frac{1}{2}(1 + \sqrt{5})$, i.e. 1.61803 to 5 decimal places.   This number, 1.618 . . . , is so interesting and important that it was given a special name by the ancient Greeks at least sixteen centuries before the time of Fibonacci.   And it seems almost certain that the Egyptians, in an even earlier epoch, ascribed magical properties to this number and used it in connection with the designs of their great pyramids. The Papyrus of Ahmes, inscribed hundreds of years before the rise of ancient Greek culture, and now in the British Museum, contains a detailed account of the

building of the Great Pyramid of Gizeh about 3070 B.C., as
culled from some much earlier writings. In this Ahmes refers to a
"sacred ratio" that was used in the proportions of that great structure: recent measurements of the pyramid itself show the ratio of
the slant edge length to the distance from ground center to base
edge as being indeed almost exactly 1.618! This ratio is the Golden
Section of the ancient Greeks.

We arrived at the Golden Section through consideration of the
Fibonacci series. But its more important property becomes apparent when we derive its value in a more direct way.

In Figure 2-5 we have a line, of length
$x + y$, marked off into two parts, x and y.
Say the respective lengths of x and y are
such that the ratio of the whole to the

FIG. 2-5.

greater part is the same as the ratio of the greater to the lesser part.

i.e.,
$$\frac{x + y}{x} = \frac{x}{y}.$$

Then $x^2 - xy - y^2 = 0$, whence $\dfrac{x}{y} = \dfrac{(1 + \sqrt{5})}{2}$.

So here we find that the ratio of x to y, that is the ratio of the
greater part to the lesser part, is that same Golden Section.

At this point we can note two rather amusing, although quite
unimportant, little facts regarding the Golden Section.

It is the only number which is transformed into its own reciprocal by merely subtracting 1!

i.e.,
$$\frac{(1 + \sqrt{5})}{2} - 1 = \frac{2}{(1 + \sqrt{5})}.$$

Then, trigonometrically, in terms of the well-known mathematical
constants e and i $(= \sqrt{-1})$, the Golden Section can be shown as:

$$2 \cos\left(\frac{\log_e (i^2)}{5i}\right).$$

Now we'll take yet another of the many approaches that lead to the
Golden Section, this time through consideration of the pentagon
which was associated with so many ancient beliefs and pagan rites.
In a regular pentagon $ABCDE$, Figure 2-6, of side p units, we have
the lines AC and BE intersecting at F, with $CF = x$ and $AF = y$.

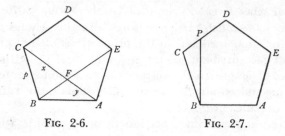

FIG. 2-6.   FIG. 2-7.

Then triangle ABC is isosceles, with $\angle ABC = 108°$, so $\angle BAC = \angle BCA = 36°$.

Also, $BF = AF$,   so $\angle ABF = \angle BAF = 36°$.

Hence,

$\angle CBF = 108° - 36° = 72°$,   and $\angle CFB = 180° - 108° = 72°$.

Thence, $CF = CB$,   so $x = p$.

But, $x + y = 2p \cos 36°$,   and $y = \dfrac{p}{2 \cos 36°}$,

so $x = \dfrac{p(4 \cos^2 36° - 1)}{2 \cos 36°}$.

Hence, $4 \cos^2 36° - 1 = 2 \cos 36°$, i.e., $4 \cos^2 36° - 2 \cos 36° - 1 = 0$, with the positive solution

$$\cos 36° = \tfrac{1}{4}(1 + \sqrt{5}).$$

Then, $x + y = \tfrac{1}{2}p(1 + \sqrt{5})$, $x = p$, $y = \dfrac{2p}{(1 + \sqrt{5})}$,

so $\dfrac{x + y}{x} = \dfrac{x}{y} = \dfrac{1}{2}(1 + \sqrt{5})$,

the Golden Section again!

In this, incidentally, we also proved another example of the Golden Section ratio in the pentagon. From the expression for $x + y$, i.e., AC, the ratio of AC to AB was shown to be $\tfrac{1}{2}(1 + \sqrt{5})$.

Indeed there are many ways in which the ratio appears in any study of the regular pentagon, although space will permit of only one final example.

Figure 2-7 shows the same pentagon $ABCDE$, with the perpendicular BP from the side AB meeting CD in P. Then it can be shown that the ratio of DP to CP is the Golden Section.

So much for what the Golden Section is. Now we must see why the ancient Greeks attached importance to the ratio.

The diagrams in Figure 2-8 are intended to represent two picture frames, of very different proportions. What would be your own reaction to either of these as ideal shapes for framing some picture? Would you consider one "too square" and the other "too narrow"? Surely you would! But what about the third as an example of aesthetically acceptable proportions?

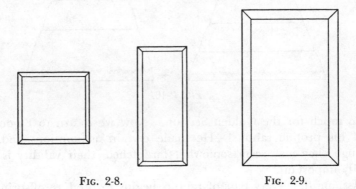

FIG. 2-8.   FIG. 2-9.

From what has been said previously, you may not be surprised to hear that this third picture frame (Figure 2-9) conforms with the Golden Section. The ratio of height to width being approximately 1.618 to 1, we have proportions that seem to satisfy the human feeling of what is right.

It was in this type of application that the Golden Section was deemed so important by architects of ancient Greece, as it has been by so many of the greatest architects and artists and sculptors right through to the present day. In almost every one of the great buildings that have survived the ravages of time and still stand as monuments to the Greek fathers of western architecture one can see example after example of the Golden Section ratio, in the proportions of façades, archways, doors, and other vital parts of the structures. In more recent times the ratio was applied in many of their greatest works by Leonardo da Vinci, Christopher Wren, and the Adams brothers. And it appears even today in many of the finest examples of contemporary architecture, art, and design—from skyscrapers to squaw-skirts! Yes, even in the more stylish of those gay skirts so well known to devotees of the square-dance!

Here we have representation of two somewhat different styles for the 4-tier squaw-skirt (Figure 2-10). Which would you prefer? Surely, the right-hand example with its tiers of graded depths. And, in that design, we have the Golden Section as the ratio of the depth of each tier to that of the tier above it.

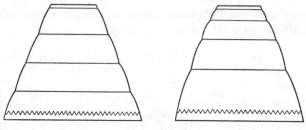

FIG. 2-10.

So much for the Golden Section. Now we return to Fibonacci and his prolific rabbits! He made certain assumptions and, although they may seem somewhat far-fetched, their validity is not really important.

Assume that every pair of rabbits begins to produce offspring at the age of two months, begetting just one pair every month thereafter, and that rabbits don't die!

Starting with only one pair, in the second month there will still be one pair, but two pairs in the third month, three pairs in the fourth, five pairs in the fifth, eight pairs in the sixth month, and so on. The number of pairs in each month, starting at the first month, will be given by the corresponding term of the Fibonacci series.

It would be a tedious chore to find the value of, say, the 30th term in the Fibonacci series: the number of pairs of rabbits in the 30th month. It would be an even more tedious task to calculate this value using the simple formula that gives it. However, the formula enables one to arrive at a good approximation when used in conjunction with logarithms.

The value of the nth term in the series is:

$$\frac{(1 + \sqrt{5})^n - (1 - \sqrt{5})^n}{2^n \sqrt{5}}$$

For example, the 50th term is exactly 12,586,269,025.

In fact, provided n is large, a reasonably accurate approximation to the value of the nth term is given by:

$$\frac{1.618^n}{2.236},$$ and note the appearance of 1.618 here.

Using this simple formula, with only 4-figure logarithms, we derive 12,600,000,000 as a fair approximation to the 50th term.

Quite apart from its connection with the Golden Section, the Fibonacci series is most interesting in itself. Many unexpected relations can be found between its terms, and these have been the subject of much study and innumerable theorems—some trivial, some extremely complex in the Theory of Numbers.

Here we show a few of the more obvious relations within the Fibonacci series. Many others of the same general type can be found by careful study of the successive terms, and a search for these may provide quite an interesting pastime.

We use the notation previously defined, the nth number in the series being shown as f_n, with $f_1 = f_2 = 1$.

$$\text{Sum of the first } n \text{ terms} = f_{n+2} - 1$$
$$\text{Sum of the first 10 terms} = 11f_7$$
$$f_n f_{n+1} - f_{n-1} f_{n-2} = f_{2n-1}$$
$$f_{n-1}^2 + f_n^2 = f_{2n-1}$$

Amongst the more complex relations there are two of special interest.

(1) It can be proved that, if f_p be a prime number ($p > 4$), then p must be a prime number. For example, $f_{11} = 89$, which is prime, so 11 is prime.

The converse of this is not necessarily true, i.e. if $p = 31$ (prime), $f_{31} = 1,346,269 = 2417 \cdot 557$.

(2) It can be proved that, if p be a prime number, then:
  if p be of form $(10k \pm 1)$, $f_p = ap + 1$;
  if p be of form $(10k \pm 3)$, $f_p = bp - 1$;
  where a and b are integers.
The converse of this is not necessarily true: i.e., $f_{14} = 377 = 27 \cdot 14 - 1$.

For example, we list the following:

$p = 10k \pm 1$	f_p	$p = 10k \pm 3$	f_p
11	89	7	13
19	4181	13	233
29	514229	17	1597
31	1346269	23	28567
41	165580141	37	24157817

But here we have been hovering on the very brink of serious mathematics. Let us draw back, and examine the intriguing manifestations of the Fibonacci series that are found in the field of botany, in familiar plants, flowers, and fruit available in our homes. These are seen in the phenomenon called *phyllotaxis,* meaning the arrangement of leaves on a stem.

Here we have part of a twig from a cherry tree. If a thread were passed from one leaf to the next, continuing from leaf to leaf in the same direction along and around the twig, a spiral would result. And one would return relatively to the initial position every fifth leaf and every second complete circuit of the twig.

The arrangement of leaves can be expressed, then, as a fraction:

$$\frac{\text{Number of complete turns}}{\text{Number of leaves per cycle}}$$

In the case of the cherry, the phyllotaxis is $\frac{2}{5}$, as it is with the oak and many other trees and plants. In some cases, for example

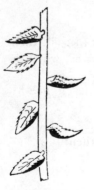

FIG. 2-11.

the elm, the leaves are found alternately on opposite sides of the stem and there the phyllotaxis is $\frac{1}{2}$. In the familiar beech, where the arrangement repeats every third leaf with one circuit to the cycle, the phyllotaxis is $\frac{1}{3}$. Other examples are the pear, $\frac{3}{8}$, and the willow, 5/13.

You will find examples galore if you look for them. In many grasses the phyllotaxis is $\frac{1}{2}$, but more often $\frac{1}{3}$. And flowers provide many examples. In the tulip, the leaves are generally in $\frac{1}{3}$ phyllotaxis. Even tulip petals, which are modified leaves, are arranged in two whorls of three and can be regarded as telescoped turns of the phyllotaxis spiral.

Indeed, so far as manifestations of the phenomenon in leaves and blossoms are concerned, it seems probable that each species and often each complete botanical 'family' conforms to a particular phyllotaxis.

The interesting point about all this is that all these phyllotaxis "fractions" are made up of alternate numbers in the Fibonacci series. Ratios which do not conform to this rule are never found in the phyllotaxis of plants, except where damage or abnormal twisting of a stem has modified the whole arrangement.

Higher ratios in the Fibonacci series are mostly found in plants with greatly shortened stems, ratios such as 13/34, 21/55, and even higher being found in many of the common rosette plants, and in a sense in the arrangements of the scales of a pine cone.

With the pine cone, the pineapple, and even the arrangement of seeds in a sunflower we have something rather different. The scales of a pine cone are really modified leaves, crowded together and in contact on a very short stem. Here we do not find phyllotaxis in the sense that applies with true leaves and suchlike. However, we can detect prominent arrangements of ascending spirals. These are the *parastichies,* resulting from direct contact of those "leaves" between which the lateral distance on the axis of the cone is the least.

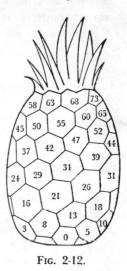

Fig. 2-12.

Figure 2-12 shows a schematic representation of a pineapple, with the scales numbered in order, this order being based on the corresponding lateral distances of the scales along the axis of the pineapple: scale 5 is slightly "higher" along the axis towards the leafy crown than scale 4, which is out of sight; similarly, scale 4 is slightly "higher" than scale 3.

Three distinct groups of parastichies are obvious. One, typified by the spiral 0, 5, 10, etc ascends at a shallow angle. The second ascends in the same direction, but very steeply, as the spiral 0, 13, 26, etc. The third, in the opposite direction, is typified by the spiral 0, 8, 16, etc.

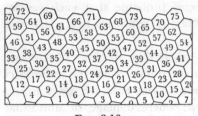

FIG. 2-13.

Now imagine that we can lay out the surface arrangement of this pineapple, still adhering to our numbering of the scales, as shown in Figure 2-13.

The three different systems of parastichies show up very clearly. Also, it may be noted that there can be only three systems with scales actually contiguous, since the pineapple scales are of irregular hexagonal shape. And the three systems provide arithmetical series, with common differences respectively 5, 8, and 13: three consecutive numbers in the Fibonacci series.

Is it not amazing that this simple series turns up in so many and so varied connections, in man's perception of what is beautiful and in the mysterious workings of nature itself?

Chapter 3

MYSTIC ARRAYS

In this chapter we are going to deal with one of the purest of the pastimes of recreational mathematicians. The construction and study of arrays of numbers having certain properties has been going on for quite a number of centuries and, until recently, a practical use has escaped all.

It is not often that an old-fashioned term remains attached to a topic that has been as thoroughly studied, mathematically, as *Magic Squares*. Literally countless books, articles and pamphlets have been written about magic squares and their mathematical and mystical properties.

For centuries magic squares seemed to defy attempts to put them to practical use (discarding their use by astrologers and mystics). In a world which seems to be predicated on "what good is it?", it is sad to say that magic squares, too, have found their niche in civilized scientific research. However, more of that later. Let us concentrate on the *delights* of magic squares.

There seems to be no known point in time at which we can say that magic squares were first noted. Legend has it that a turtle was found with a magic square on its shell by a Chinese emperor centuries before the birth of Christ. The *Lo Shu* magic square shown in Figure 3-1 (originally in non-arabic numerals, of course) was known about 1000 B.C.

This particular square has a long history of magic and mystery in China where its use as a charm or talisman is common. Shuffleboard players will recognize the pattern.

The magic constant of this square is 15, i.e., the sum of each of the three rows, three columns and two main diagonals is 15. It will also be seen that only the integers from 1 to 9 are used in the nine squares, or *cells*. Many other 3rd-order (3 by 3) magic squares can be constructed using other numbers but, unless some special cases arise, we shall deal only with *normal* magic squares having the integers from 1 to n^2 for nth-order magic squares.

Fourth-order magic squares are next in line and some general properties and definitions can be made clear by using several of them as examples.

Figure 3-2 shows one of the most famous magic squares. Albrecht Dürer used this one in his engraving called *Melancholie* which depicts the indecision of the intellectual. A disorderly array of the instruments used in science and construction is shown unused while the thinker sits deep in thought. Dürer shows the magic square on a column and the middle two numbers, 15, 14, in the bottom row date the engraving.

The magic constant of Dürer's magic square—and of all normal 4th-order magic squares—is 34. That the rows, columns and two main diagonals each total 34 is sufficient to classify the figure as a normal magic square. Figure 3-3 shows what is called a *diabolic*, or panmagic or pandiagonal, magic square. This square is also

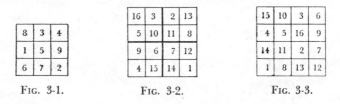

FIG. 3-1. FIG. 3-2. FIG. 3-3.

magic along the broken diagonals (e.g. 15 + 8 + 2 + 9, 4 + 10 + 13 + 7 and others). Dürer's magic square has another property in that there are five groups of four cells which total 34—the four quadrants marked off in bold lines and the middle four cells. Figure 3-3 has this property with a bonus: *any* 2 by 2 arrangement here totals 34. Then, thinking of such an arrangement as the 15-10-1-8 combination as a 2 by 2 "by continuation," one can find more 34's.

The magic constant of any normal magic square of order n is

$$\frac{n(n^2 + 1)}{2}$$

So a 4th-order magic square has a constant of 34, a 5th-order magic square has a constant of 65, and so on.

There are 880 different 4th-order magic squares, not counting those obtained by rotation or reflection. Of these, there are 384 diabolic magic squares. It still remains one of the unsolved problems in recreational mathematics exactly how many different magic squares of order n exist. There are no 2nd-order magic squares (unless we use four ones—which doesn't follow the "rules of the game") and there is only one basic 3rd-order magic square—though eight patterns can be produced by appropriate rotation and reflection of the basic pattern. The number of 4th-order magic squares has been given and an idea of the tremendous gain in possibilities may be seen from the fact that the number of different 5th-order magic squares is estimated (it is not known exactly) to be greater than 13,000,000. We do know that there are exactly 3600 different 5th-order diabolic magic squares (28,800 if one includes rotations and reflections). All of these 3600 diabolic squares have been tabulated by Francis L. Miksa, who has worked out a linear transformation method to find them from the four truly basic nonequivalent diabolic magic squares shown in Figure 3-4.

18	24	5	6	12
10	11	17	23	4
22	3	9	15	16
14	20	21	2	8
1	7	13	19	25

15	18	21	4	7
24	2	10	13	16
8	11	19	22	5
17	25	3	6	14
1	9	12	20	23

20	22	4	6	13
9	11	18	25	2
23	5	7	14	16
12	19	21	3	10
1	8	15	17	24

13	17	21	5	9
25	4	8	12	16
7	11	20	24	3
19	23	2	6	15
1	10	14	18	22

FIG. 3-4.

Here we shall give a few simple methods of constructing magic squares of any order. Throughout this chapter, the individual little squares within the magic squares will be referred to as "cells." The easiest method known can be used to construct odd-order magic squares. Examine the 5th-order magic square in Figure 3-5 and then we shall use it to illustrate the general method of construction.

Note that in most cases, successive integers are arranged in a diagonal relationship. The method of construction, called the de la Loubère method, shown in Figure 3-6, is to start with 1 in the top middle cell and continue to write successive integers following a right-upward diagonal path. When the next number would fall outside the confines of the array, we merely imagine the column to begin at the directly opposite point. For example, 2 would fall in the imaginary cell to the upper right of 1. If we imagine the plane to be covered with 5 by 5 arrays, this position of 2 corresponds to the fourth cell of the bottom row. This is where we place the number 2. 3 follows the rule and 4 would then carry over across the array to the position shown. We note, as we continue, that we have no place to put 6, since 1 occupies the next cell. All we do when we run up against a previously occupied cell is to drop one cell down and then continue following the rule. After reaching 15 we would run off the square into the *corresponding* lower left corner *which is already occupied by 11:* so we drop down to the cell below 15. A couple of checks that can be made to ensure that we haven't erred in applying the rule are to see that the middle cell is occupied by the $[(n^2 + 1)/2]$th term (13 in the case of the 5th-order magic square shown), and that the last member of the series, n^2, falls in the middle cell of the bottom row. We have constructed an ordinary, not diabolic, magic square.

17	24	1	8	15
23	5	7	14	16
4	6	13	20	22
10	12	19	21	3
11	18	25	2	9

FIG. 3-5.

	18	25	2	9	11	
17	24	1	8	15	17
23	5	7	14	16	23
4	6	13	20	22	4
10	12	19	21	3	10
11	18	25	2	9

FIG. 3-6.

65	86	17	38	59
83	29	35	56	62
26	32	53	74	80
44	50	71	77	23
47	68	89	20	41

FIG. 3-7.

This method by de la Loubère will work for any odd-order square that comprises integers in arithmetic series. Following the rules, we can construct such squares as the one in Figure 3-7. The series of integers starts with 17 and continues with every third integer (17, 20, 23, 26, etc.). The constant for this square is 265.

The magic constant for any n-order magic square starting with the integer A, with an arithmetic series with a common difference of D, has a constant equal to

$$n \left[\frac{2A + D(n^2 - 1)}{2} \right]$$

For a normal magic square $A = D = 1$ and the above formula reduces to the one shown on page 24.

Another method of constructing odd-order magic squares was devised by Bachet de Méziriac and is very similar to de la Loubère's method just described.   The number 1 goes in the cell above the middle cell and then de la Loubère's diagonal rule is followed throughout until we get to an occupied cell.   We then continue in the same column two squares higher.   Fig-

46	15	40	9	34	3	28
21	39	8	33	2	27	45
38	14	32	1	26	44	20
13	31	7	25	43	19	37
30	6	24	49	18	36	12
5	23	48	17	42	11	29
22	47	16	41	10	35	4

FIG. 3-8.

ure 3-8 shows a 7th-order magic square constructed by de Méziriac's method.   Note that the S.W./N.E. diagonal is the same as is generated by de la Loubère's method.

The construction of even-order magic squares has always presented difficulties and we shall describe a general method and two specific methods of construction.

The general method to be described is credited to de la Hire (the observant reader will probably note that the French seem to have been particularly intrigued by magic squares).   We shall construct a 4th-order magic square by de la Hire's method, indicating how the method can be used to construct any even-order magic square.

First write the numbers 1 to 4 in order in the two main diagonals as shown in Figure 3-9a and then use the same numbers so that the rows and columns equal 10 as shown in Figure 3-9b.   (For any n-order magic square —n even—we would write the numbers 1 to n in the two main diagonals of an n by n array.)   Now reverse the row-column position of 3-9b to form Figure 3-9c.   The numbers which appear in Figures 3-9b and 3-9c we shall refer to as *primary* numbers.

Form a *root* square, Figure 3-10a, from 3-9c by substituting root numbers for the primary numbers (see table alongside Figures 3-10) and then add the corresponding numbers of Figure 3-9b to those in Figure 3-10a to give the 4th-order magic square shown in Figure 3-10b with a magic constant of 34.

1			4
	2	3	
	2	3	
1			4

(a)

1	3	2	4
4	2	3	1
4	2	3	1
1	3	2	4

(b)

1	4	4	1
3	2	2	3
2	3	3	2
4	1	1	4

(c)

FIG. 3-9.

Primary	Root
Numbers	*Numbers*
1	0
2	4
3	8
4	12

0	12	12	0
8	4	4	8
4	8	8	4
12	0	0	12

(a)

1	15	14	4
12	6	7	9
8	10	11	5
13	3	2	16

(b)

Fig. 3-10.

For any even-order magic square the root numbers can be calculated from the corresponding primary numbers:

Root Number $= n(p - 1)$, where n denotes the order (even), and p the Primary Number ($p = 1, 2, 3$, etc., up to n).

e.g.,

6th Order: Primary Numbers: 1   2   3   4   5   6
          Root Numbers: 0   6 12 18 24 30

10th Order: Primary Numbers: 1   2   3   4   5   6   7   8   9 10
           Root Numbers: 0 10 20 30 40 50 60 70 80 90

In one of the initial steps above we supplied primary numbers so that the rows and columns would have the same sum. For the 4th-order square we constructed we used 10 as the sum and this is nothing more than the sum of the numbers 1, 2, 3, 4. For n-order magic squares this initial sum is always $1 + 2 + 3 + 4 + \ldots + n = [n(n + 1)/2]$.

Figure 3-11 shows an 8th-order magic square constructed by the de la Hire method. Starting with two main diagonals using the numbers 1 to 8, then using these digits to fill in the remaining cells so that the rows and columns sum

1	63	62	4	5	59	58	8
56	10	11	53	52	14	15	49
48	18	19	45	44	22	23	41
25	39	38	28	29	35	34	32
33	31	30	36	37	27	26	40
24	42	43	21	20	46	47	17
16	50	51	13	12	54	55	9
57	7	6	60	61	3	2	64

Fig. 3-11.

to 36, and following through with the general procedure outlined above. The table needed in this case:

8th Order: Primary Numbers: 1   2   3   4   5   6   7   8
          Root Numbers: 0   8 16 24 32 40 48 56

There are two relatively simple methods by which we can construct the two basic types of even-order magic squares, viz., where

$n = 4m + 2$ (e.g., $n = 6$, 10, 14, 18, etc.) or where $n = 4m$ (e.g., $n = 4$, 8, 12, 16, etc.). These are, respectively, the singly-even and the doubly-even magic squares.

We can construct a singly-even magic square by using Ralph Strachey's method. A 6th-order magic square gives us $6 = 4m + 2$ where $m = 1$. Form four quarters as in Figure 3-12 and in each of the quarters form a 3rd-order magic square using the digits 1 through 36 (1 to 9 in square A, 10 to 18 in B, 19 to 27 in C, and 28 to 36 in D) by using de la Loubère's method. In the middle row of A take the m cells after the first one and in each of the other rows

A	C
D	B

FIG. 3-12.

8	1	6	26	19	24
3	5	7	21	23	25
4	9	2	22	27	20
35	28	33	17	10	15
30	32	34	12	14	16
31	36	29	13	18	11

FIG. 3-13.

35	1	6	26	19	24
3	32	7	21	23	25
31	9	2	22	27	20
8	28	33	17	10	15
30	5	34	12	14	16
4	36	29	13	18	11

FIG. 3-14.

of A take the m cells from the left edge and interchange these cells with the corresponding cells in D. Now interchange the cells in the $m - 1$ columns next to the right edge of C with the corresponding cells of B. Figure 3-14 results which is a 6th-order magic square. The last column is not different from the initial square of Figure 3-13 because the interchange of $m - 1$ columns means interchanging zero columns. However, the 10th-order magic square of Figure 3-15 shows that interchanges take place along the right edges of C and B.

92	99	1	8	15	67	74	51	58	40
98	80	7	14	16	73	55	57	64	41
4	81	88	20	22	54	56	63	70	47
85	87	19	21	3	60	62	69	71	28
86	93	25	2	9	61	68	75	52	34
17	24	76	83	90	42	49	26	33	65
23	5	82	89	91	48	30	32	39	66
79	6	13	95	97	29	31	38	45	72
10	12	94	96	78	35	37	44	46	53
11	18	100	77	84	36	43	50	27	59

FIG. 3-15.

Construction of doubly-even magic squares is even simpler. Merely construct, in the case of a 4th-order square, the 16-cell array writing in the digits from 1 to 16 in order across the rows as shown in Figure 3-16. Now simply switch diagonal pairs of numbers as shown in Figure 3-17—which is a 4th-order magic square. To construct larger doubly-even magic

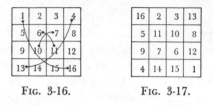

FIG. 3-16.   FIG. 3-17.

squares use the same technique. For an 8th-order magic square, write the numbers 1 to 64 in order along the rows and perform the diagonal switches indicated (only one-half of the switches are shown

1	2	3	4	5	6	7	8
9	10	11	12	13	14	15	16
17	18	19	20	21	22	23	24
25	26	27	28	29	30	31	32
33	34	35	36	37	38	39	40
41	42	43	44	45	46	47	48
49	50	51	52	53	54	55	56
57	58	59	60	61	62	63	64

64	2	3	61	60	6	7	57
9	55	54	12	13	51	50	16
17	47	46	20	21	43	42	24
40	26	27	37	36	30	31	33
32	34	35	29	28	38	39	25
41	23	22	44	45	19	18	48
49	15	14	52	53	11	10	56
8	58	59	5	4	62	63	1

FIG. 3-18.   FIG. 3-19.

to avoid a mess of lines—similar switches are made along the other main diagonal and horizontal mid-section). The resulting 8th-order magic square is shown in Figure 3-19.

Magic squares of all kinds can be constructed including those which are magic for multiplication. Figure 3-20 shows such a magic square in which the product of the numbers in any row or column or main diagonal is equal to 4096. This was easily constructed by starting with a 3rd-order magic square using the integers from 0 to 8 and then substituting the corresponding powers of 2: corresponding powers of any other constants could have been used to provide multiplication magic squares.

128	1	32
4	16	64
8	256	2

FIG. 3-20.

Addition-multiplication magic squares are known (but not so easily constructed) which are magic not only for addition but for multiplication as well. Figure 3-21 shows a magic square which has an addition magic constant of 840 and a multiplication magic constant of 2,058,068,231,856,000. The reader may verify this—if he desires!

46	81	117	102	15	76	200	203
19	60	232	175	54	69	153	78
216	161	17	52	171	90	58	75
135	114	50	87	184	189	13	68
150	261	45	38	91	136	92	27
119	104	108	23	174	225	57	30
116	25	133	120	51	26	162	207
39	34	138	243	100	29	105	152

FIG. 3-21.

Magic cubes built up of stacks of magic squares are also known and can be constructed by standard procedures. Figure 3-22 shows a 3rd-order magic cube shown in perspective and also broken down into the component 3rd-order magic squares. The magic constant of 42 is found along all 27 rows and columns and along the four main diagonals of the cube.

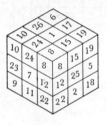

10	24	8
23	7	12
9	11	22

26	1	15
3	14	25
13	27	2

6	17	19
16	21	5
20	4	18

FIG. 3-22.

A most interesting class of magic squares comprises multi-magic squares. These are squares which are magic in the usual way and also if the whole set of numbers is raised to the 2nd or higher power. Figure 3-23 shows a doubly magic square which has a constant of 260 for addition and a constant of 11180 after the numbers are squared and then added.

Trebly-magic squares are known which remain magic when squared also when raised to the third power. However, the smallest known magic square fulfilling these conditions is a 64th-order square.

16	41	36	5	27	62	55	18
26	63	54	19	13	44	33	8
1	40	45	12	22	51	58	31
23	50	59	30	4	37	48	9
38	3	10	47	49	24	29	60
52	21	32	57	39	2	11	46
43	14	7	34	64	25	20	53
61	28	17	56	42	15	6	35

FIG. 3-23.

Let us temporarily leave the realm of mathematical arrays and devote a little attention to examples of magic *word* squares. These, too, have interested mathematicians as a sort of "break" from number magic squares.

Crossword puzzles may be called a primitive form of magic squares of words. However, it is not so easy to form arrays of words in an *n*-order array such that all the cells are occupied by letters and all the rows and columns form words.

Here is a small teasing collection of word squares—we omit the methods of construction (for even here, there are mathematical methods of construction!).

G	A	P
A	R	E
P	E	T

R	A	R	E
A	V	I	D
R	I	S	E
E	D	E	N

B	I	S	H	O	P
I	M	P	A	L	E
S	P	I	N	E	T
H	A	N	G	A	R
O	L	E	A	T	E
P	E	T	R	E	L

Fɪɢ. 3-24.

S	A	T	O	R
A	R	E	P	O
T	E	N	E	T
O	P	E	R	A
R	O	T	A	S

Sator, arepo tenet opera rotas

(Arepo, the sower, delays the wheels by his works.)

T	O	T	I
E	M	U	L
E	S	T	O

Fɪɢ. 3-25.   Fɪɢ. 3-26.

And an old word square in Latin which also forms a sentence is shown in Figure 3-25. A mysterious inscription appears on a pillar in Texas and we've reproduced it in Figure 3-26.

A startling trick can be played with another kind of *"magic"* square. Ask someone to pick out any number in the array shown in Figure 3-27 and then cross off all the horizontal and vertical numbers in line with this number. Then have him pick another number that has not been crossed off or eliminated and repeat the elimination procedure. After five numbers have been chosen, there will be no more available. The sum of the chosen numbers will always total 666 (the number of the beast). Try it.

125	191	248	169	116
48	114	171	92	39
136	202	259	180	127
69	135	192	113	60
64	130	187	108	55

FIG. 3-27.

	17	83	140	61	8
108					
31					
119					
52					
47					

FIG. 3-28.

Although the numbers appear to be randomly arranged (and many other squares can be constructed which force 666—or any other number) the trick is surprisingly simple.

666 can be broken down as the sum of 10 numbers in thousands of different ways. Take any one of these, say for example, the summation of:

$$8, 17, 31, 47, 52, 61, 83, 108, 119, 140$$

Write the numbers, in any order, outside the borders of a 5 by 5 grid as shown in Figure 3-28. Now fill the cells, each being given the sum of its corresponding row and column indicators (e.g., 31 + 140 = 171). The process described in the trick does nothing more than force the addition of the 10 numbers with sum 666. Obviously, the same method can be used to construct arrays with which other selected numbers can be forced.

But are magic squares good for nothing but mathematical fun and making tricks? By no means.

Agricultural research has been furthered by the application of certain types of magic squares called Latin Squares. R. A. Fisher originated the use of Graeco-Latin squares in agricultural experimentation at the Rothamsted Experimental Station in England. Two Latin squares are shown below using only the numbers 1 to 4 in each array and having a magic constant of 10 for the rows and columns (the construction of even-order magic squares was out-

1	2	3	4
2	1	4	3
3	4	1	2
4	3	2	1

1	2	3	4
3	4	1	2
4	3	2	1
2	1	4	3

FIG. 3-29.

11	22	33	44
23	14	41	32
34	43	12	21
42	31	24	13

FIG. 3-30.

lined on page 27, using Latin squares as starting points). Now combine both of these Latin squares to form what is called a Graeco-Latin square as shown in Figure 3-30.

Suppose we wished to test the effects of four fertilizers, 1, 2, 3, 4, on the growth of wheat in a field having the usual irregularities of soil fertility. By planting the wheat and distributing the fertilizers according to a Latin square arrangement, a statistical analysis can easily determine the effect of each of the fertilizers while eliminating the effects of the different soil properties. By using the Graeco-Latin square we could also test the effects of these four fertilizers on four different brands of wheat, 1, 2, 3, 4. Distribute the fertilizers according to one Latin square and the wheat according to the other Latin square—forming the Graeco-Latin square of fertilizer-wheat patterns. The field will then be planted according to the plan of Figure 3-30. Each wheat is planted against each fertilizer and all the combinations are scattered randomly over the field. Obviously, using Latin squares of higher orders* will further minimize the effect of soil variation. Then statistical analysis can determine the comparative effect of each fertilizer on each wheat.

More recently, atomic research has benefited from magic square studies. By arranging materials being studied in atomic piles according to certain magic square arrangements, the time required for a given study can be considerably reduced without significantly affecting the value of the results obtained.

Even in such fields as marketing research and sociology there are valuable applications of Graeco-Latin squares.

Much has been written, and there remains much to be discovered regarding magic squares and other arrays of numbers in this great branch of recreational mathematics. This chapter may have served as an introduction to some of their variations and applications.

* The numbers of Latin squares of orders up to order 7 are tabulated below:

n-Order	N
2	2
3	12
4	576
5	161,280
6	812,851,200
7	61,428,210,278,400

Chapter 4

TOPOLOGICAL DELIGHTS

Seeing is believing, and there are many aspects of mathematics that leave the uninitiated in a state of complete disbelief. The theory of relativity is one; the miracle of the electronic computer, studies of the fourth and even higher dimensions, and topology are others.

Einstein's Theory of Relativity has been confirmed in most of its conclusions by physical experiments which, in the enormity of the difficulties to be overcome in measuring the almost immeasurable, are as incredible as some of the logical conclusions in the Theory itself. Computers remember, and some believe they actually think: but to most people they remain a totally incomprehensible mystery. As for the fourth dimension, even mathematicians admit they cannot visualize it!

What about topology? It has been called "rubber-sheet" geometry, and this may give a clue to its nature. Firstly, however, we may say that topology includes the study of such improbable oddities as one-sided sheets of paper and closed bottles with no insides! Can such objects be constructed? They can be, and through their construction we may begin to realize that other impossible-sounding concepts could well turn out to be quite comprehensible.

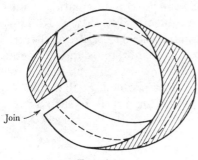

Join

Fig. 4-1.

Take a long narrow strip of paper, make a half-twist and glue the ends together, forming the band shown in Figure 4-1. If you trace along the middle with a pencil (as indicated by the dotted line) you will eventually come back to where you started—which is hardly surprising. But you will find that "both" sides of the band have been marked

35

with the pencil! And without going over an edge. This is the Möbius band, a truly one-sided surface: furthermore, it has only one edge!

This remarkable object was named after the great German mathematician, Augustus Möbius, who is generally credited with having invented it about 1855. But this very phenomenon was the subject of a discourse by Johann Benedict Listing, written in 1847.

The closed bottle with no "inside" is illustrated in Figure 4-2. A tube of material is bent, and one end is passed through its wall. The ends of the tube are then joined together, forming a continuous closed surface. Normally, if the continuous closed surface of a 3-dimensional object is penetrated, one has to repenetrate to escape. Not so with the Klein bottle. We can penetrate the wall of the bottle from the outside, and escape by merely travelling *inside* the bottle to the "opening"—escaping without repenetration of the wall.

So here we have demonstrated two cases of the "impossible"! No use has yet been found for the Klein bottle: liquids would spill from it rather easily. But B. F. Goodrich Company has patented the use of the Möbius band as a conveyor or machinery belt: "both" sides being the one side, the belt lasts twice as long as a conventional belt.

The theoretical aspects of topology involve concepts of enormous complexity. In fact, there are many problems in this field that have not yet been solved. However, its more practical aspects are well within the understanding of young children, who even enjoy their excursions into this challenging field.

Leonard Euler, who contributed so much to mathematics, made the first true study of a topological problem. The city of Königsberg (now known as Kalingrad) is situated astride the Preger River, and includes the island of Kneiphof and another island. There were seven bridges connecting the parts of the city, as shown in Figure 4-3 (there are now 8 bridges). The problem at that time—the 18th century—was the possibility of taking a walk in such a way

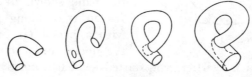

FIG. 4-2.

FIG. 4-3.

as to cross each of the seven bridges once, and only once, and return
to the starting point.

Euler translated the problem into the equivalent diagram also
shown in Figure 4-3. From this diagram he was able to establish
the impossibility of solving the problem under the given conditions:
these included remaining within the city limits, for obviously by
walking to the source one could reach "the other side" on dry land.
Each vertex in the diagram corresponds to one of the land masses,
and the lines correspond to the bridges.

The general rule that Euler established is that a closed network
of lines can be traced, without any retracing, only if there are
either none or two odd-order vertices in the network. An odd-
order vertex is one which has an odd number of lines radiating
from it. In the Königsberg bridges network we see four odd-order
vertices, hence that network cannot be traced without retracing
some part of it.

Figure 4-4 has no odd-order vertices, Figure 4-5 has two so these
networks can be drawn without lifting the pencil and without retracing any parts of them (they can also be drawn without crossing
any parts). How about the designs shown in Figures 4-6 and 4-7?

Euler also discovered that for all closed networks $V - L + R = 2$,
where V = the number of vertices, L = the number of lines between
the vertices, R = the number of separate regions (one region being
the whole infinity of the plane outside the network). In Figure

FIG. 4-4.   FIG. 4-5.   FIG. 4-6.   FIG. 4-7.

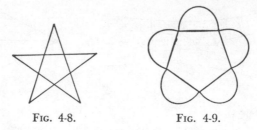

FIG. 4-8.   FIG. 4-9.

4-8, $V = 10$, $L = 15$, $R = 7$ and $10 - 15 + 7 = 2$. In Figure 4-9, which is very similar, we have $V = 5$, $L = 10$, $R = 7$ and $5 - 10 + 7 = 2$.

It is remarkable that the same formula applies when we pass from 2-dimensional plane networks to 3-dimensional solids. For any polyhedron $V - E + F = 2$, where $V =$ the number of vertices, $E =$ the number of edges connecting vertices, $F =$ the number of faces. Figure 4-10 shows three regular polyhedra with which you will be able to check the formula.

Tetrahedron   Cube   Octahedron

FIG. 4-10.

Euler's rule regarding the tracing of a closed plane network applies equally when tracing the edges of a polyhedron. The cube has 8 odd-order vertices, so it cannot be traced without retracing. The octahedron, on the other hand, can be drawn without lifting the pencil.

So far we have touched on only some topological considerations in connection with one-sided configurations: network patterns in two and three dimensions. Topology, however, embraces all the unchanging properties of geometrical figures, properties which remain essentially unaffected by deformations that may radically change various metric attributes and even the appearance of those figures.

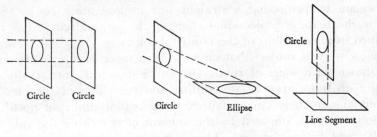

FIG. 4-11.

Projecting a circle onto a plane we obtain a circle, or an ellipse, or a straight line segment, depending on the inclination of the plane containing the original circle to the plane containing the projection.

In Figure 4-11 we see the applications of parallel line projection. But projecting with non-parallel lines as with converging or diverging beams of light, we obtain the same shaped results correspondingly scaled down or enlarged. If, however, we use non-parallel lines that are not symmetrical so far as the center of the original circle is concerned, we obtain the complete range of conic sections including the parabolas and hyperbolas, and from a circle of infinite radius to a line of infinite length. See Figure 4-12.

What property has been retained through all these quite drastic projectional changes? The original circle, and all the projections (including the straight line) are in the family of conics. A square is not in this family. We can project a square, or any plane figure for that matter, to produce a straight line. But we cannot obtain

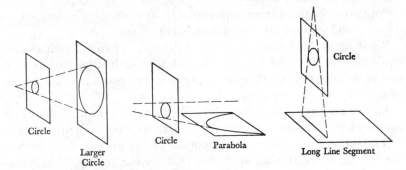

FIG. 4-12.

a square by projecting a straight line or any other conic. So, through all these projectional changes, the original circle has retained its membership in the family of conics.

Now we will subject that circle to even more drastic treatment. If drawn on a sheet of rubber it can be deformed into a square quite easily by appropriate pulling and stretching of the elastic material. In these circumstances, the sole limitations on possible deformation are imposed by the amount of stretching the rubber will stand before tearing. Then what original property of that circle will survive such deformation? There is just one property: the resulting square, or any other figure produced by stretching the material, still has an inside and an outside, as did the original circle.

This may seem a trivial deduction, but its implications have far-reaching consequences. It enables us to justify the study of a great variety of figures under some very simple assumptions and the minimum of rules. Squares, triangles, circles, ellipses, state boundaries, the gamut of regular and irregular polygons, indeed all figures with non-intersecting edges and with both inside and outside—they can all be considered as one type of configuration. Topologically, they are all equivalent.

To study equivalent objects in three dimensions we must go from rubber sheeting to wads of clay or plasticine. A ball of clay can be deformed into a cube, or any of the normal polyhedra—symmetrical or otherwise. But here again we have a limitation on possible deformation: we may compress, pull, or stretch the clay to any degree short of rupture, but we are not permitted to break the surface by poking a hole through it. Otherwise we could deform the original ball of clay to obtain a torus (i.e., a "doughnut"), or a cup, or a ring, or any other configuration topologically equivalent to a doughnut. Toruses and spheres are not topologically equivalent.

Plane figures and solid objects are defined as being topologically equivalent if they can be transformed into each other under any sort of deformation short of actual tearing or breaking of the plane or space in which they exist.

Equivalence may sometimes be difficult to determine. Examine the two knots shown in Figure 4-13. Are they equivalent?

This can be answered from a consideration of how the knot on the left could be transformed into that on the right. This transformation would obviously require cutting of the rope: of course

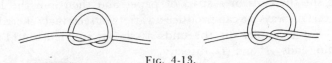

Fig. 4-13.

we would not be permitted to untie the knot, that being tantamount
to piercing or disrupting the clay in our previous example of the
clay ball. So these two knots are not equivalent.

The most famous topological problem is probably the four-color
map theorem, first mentioned by Möbius when lecturing in 1840.
This states that no map on any plane or sphere requires more than
four colors to distinguish the states from each other: of course, states
bordering each other must have different colors, but if two states
meet only at a point (as do Utah and New Mexico in the U.S.A.)
this is not deemed bordering. Nobody *had* ever been able to draw
a map that required five colors, but until recently all attempts to
prove the theorem had failed. Then, in 1976 W. Haken and K.
Appel achieved what certainly appears to be a valid proof. We use
the word "appears" because their proof depends on a combination
of theory and computer computation, rigorous on the face of it,
but a new concept in mathematics.

Now let us turn back from the approaches to serious mathematics!
Let us return to those remarkable Möbius bands and have some real
fun with them.   No mathematics will be involved: only large sheets
of strong paper, scotch tape, scissors, and some patience.

Take a Möbius band formed as shown in Figure 4-1, and cut
right along the middle of the band.   The result may be unexpected
—cutting a Möbius band in half divides it into *one* piece.   And
this is not by any means the only remarkable effect that can be
achieved.   Make another Möbius band and cut right around as
before, but this time keeping the cut about one-third of the width
from one edge.   You will cut around twice and, when you come
back to where you started, you will have divided the band into
two pieces—remarkable pieces, as you will see when you do it.

Giving two, three, or more twists to the paper before joining the
ends, and then cutting along the middle or along the one-third way
line, you will obtain other odd results.   To further complicate
matters, try cutting some of the secondary rings formed by the
original cuttings.

If we slit the ends of a strip of paper and then join the loose ends in certain ways we can produce a variety of seemingly incredible effects. For example, slit the ends as shown in Figure 4-14 and then join ends *A* and *D* directly. Pass *B* under *A* and join to *E*. Now pass *C* over *B* and under *A;* pass *F* over *D* and under *E*. Join ends *C* and *F*. If you complete both cuts around the band you'll have an interesting arrangement of three rings.

Or try an effect developed by Ellis Stanyon in 1930. Start with the strip in Figure 4-14 again. Turn

Fig. 4-14.

over *E* to the right and join to *C;* turn over *F* to the right and join to *B;* finally pass *A* under *B* and join to *D* without turning over. Cut along the two slits and see what you get.

By now, you should have a table top full of rings and bands. And, by now, you should have noticed that it takes one cut to make two rings and two cuts to make three rings. If you want four rings you need three cuts, and so on.

Maxey Brooke and J. S. Madachy have developed a technique whereby you can produce any number of rings in a chain with *only one cut!* The secret lies in first folding the paper strip lengthwise, before slitting and joining the loose ends. Begin by folding a strip of paper once, twisting the folded strip through 360° and pasting the ends together as in Figure 4-15. Three interlocking rings are produced when the band is cut all the way around through the double thickness.

Or, try *not* twisting the folded strip, but slit the ends as shown

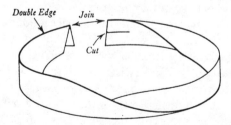

Fig. 4-15.

FIG. 4-16.

in Figure 4-16. Joining the ends as in Figure 4-17, a chain of three rings is produced by cutting the slit around the band, again through the double thickness.

For the rest of the work you'll need wider strips of paper so that the multiple folding will still leave some working space. For the 5-ring chain, to be described next, a paper strip at least four inches wide and about fifteen to twenty inches long will be needed.

A 5-ring chain can be formed by accordion-folding a paper strip as shown in Figure 4-18, slitting the ends, forming a circle and joining the ends as follows: join ends C to C, D to D, and E to E—all directly, without passing over or under any rings or ends. Now take one end of B and pass it under both of the rings C and D and join to the other end of B. Take one end of A, pass it under rings C and E and join to the other end of A. Finishing the cut (through all the folds) along the middle of the band, you will have a 5-ring chain.

If you study the general technique indicated above you have the fundamental processes involved in the formation of paper chains of any numbers of rings—all formed with a single cut after proper preparation.

Before we finish this little excursion into Ph.D.-level paper-doll cutting, let's make a 9-ring chain. We'll need a paper strip about eight to ten inches wide and about two feet long. A piece of news-

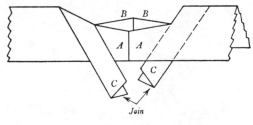

Join

FIG. 4-17.

FIG. 4-18.

paper or, for strength, the cover from a large magazine will do excellently.

Prepare an accordion-fold of seven folds as shown above in Figure 4-19, slitting the ends and marking as indicated. Join ends F to F, G to G, H to H, and I to I, directly, without passing over or under any rings or ends. Pass one end of E under rings G and I and join to the other end of E. Pass one end of D under the rings F and H and join to the other end of D. Now pass one end of C under ring G and join to the other end of C. Pass one end of A over ring G and under ring C and then join to the other end of A.

Now at this point of the process, you'll see that the ends of B are still left over. One side of the band, if cut along the middle, would fall out separately to form a 5-ring chain and the other side would fall out to form a 3-ring chain. Strip B would fall out loosely. However, we will use strip B to join the 5-ring and 3-ring chains. Ring A is an end-ring in the 5-ring chain and ring F is an end-ring in the 3-ring chain. Pass one end of B under ring A, over ring D, under ring F and then join to the other end of B. The

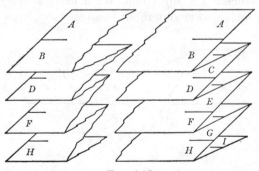

FIG. 4-19.

two end-rings, *A* and *F*, have now been joined by ring *B* and the result, maybe surprisingly, will form a neat 9-ring chain after the cut is completed around the band and through the seven folds. The rings should be in the order *H, D, F, B, A, C, G, E, I*.

If you failed to follow the directions exactly, you will end up with something entirely different. In fact, much of the fun is in the surprises that can result from deliberate and intentional mistakes!

And finally some problems for your amusement in this field. Each can be solved using one cut on a properly folded, slit, and joined band:

(1) A 4- ring chain.

(2) Two separate chains of 2 and 3 rings each.

(3) Three rings in a chain with a fourth ring looped through the middle ring.

(4) Three separate chains of 3 rings each.

(5) Two separate chains of 4 and 5 rings each.

Highly complex topological theory is bound up with many of the tricks and illusions with which magicians and conjurers mystify their audiences. Fortunately, no comprehension of such theoretical concepts is needed for the understanding of the tricks. So it may be appropriate to conclude this chapter with the outline of a few simple tricks which demonstrate yet again that the apparently impossible is often very possible when topology is involved.

Prepare a pencil and string as shown in Figure 4-20, the *loop* of string being definitely shorter than the pencil. Then thread the pencil through a friend's buttonhole, but in such a way that the string will be threaded through its own loop as shown in Figure 4-21. Of course, you are not allowed to break, untie, or remove the string from the pencil. "Impossible!" you say? By no means so, for *it can be done:* merely an application of topological theory.

Then there's the party trick that can only upset the equanimity of the two victims. Two strong cords are needed for this, the ends of one being tied to the wrists of one victim, the ends of the other cord to the wrists of the other victim—but with the strings interlocked. See Figure 4-22. The two are then invited to separate without cutting or untying the cords or removing them from their wrists. If the right pair have been picked, their fruitless efforts

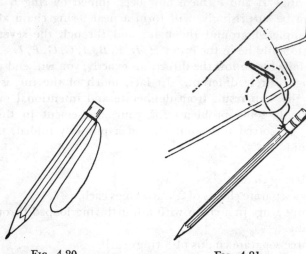

FIG. 4-20.   FIG. 4-21.

should be most amusing! And this, again, is an example of topological theory put to "practical" use.

Topology, however, is far from being merely a source of tricks.
In the case of the Möbius band a trick, based on theory, has led to
a very valuable practical application in industry. There are also

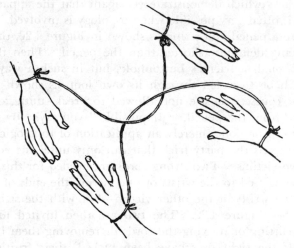

FIG. 4-22.

cases where an initial trick has led to real advances in topological theory, which have in turn touched off important developments in other branches of mathematics and science.

In Chapter 7 there will be a brief return to topology in connection with another important property associated with the deformation of objects. And, in the meantime, we hope this short excursion into the realm of topology will not have left you in a state of complete disbelief.

Chapter 5

SOME INFERENTIAL PROBLEMS

Many "inferential problems"—problems in logic—may be solved very neatly by the methods of elementary Boolean Algebra, which is the basis of the mathematics of logic. We give only a bare outline of the simplest aspects of this here, but it may encourage the reader to delve more deeply into a field in serious mathematics which has rarely been touched in its recreational applications.

We adopt the convention that something "true" has the value of 1, and something "false" the value of 0 (i.e., zero). Using code symbols for the "somethings," we can then form expressions and equations based on the known facts, and these may then be treated very much the same as those in normal algebra. A very simple example will show how these ideas are applied.

Say we have two conflicting statements about the name of a boy, and we know that each statement contains one mistake. One said "Jack Dibble," the other "John Dibble."

Obviously, Dibble was his name, and his first name was neither Jack nor John. But let us see how this would be handled by the use of Boolean Algebra.

We have only the two numerical values, 0 and 1. Nothing can be more true than "true": if, in the course of the working, we derive **any number greater than unity we can represent it as unity.**

Let A stand for Jack, B for John, and C for Dibble. Then each statement can be represented in two ways:

Multiplication—If both A and C were equal to 1 (i.e., true), the product $AC = 1$. But if either A or C has the value 0 (i.e., false), then $AC = 0$.

Addition—If either A or C (or both) has the value 1 (i.e., true), then $A + C = 1$.

Now, from the two statements, $A + C = 1$, and $B + C = 1$.

So $$(A + C)(B + C) = 1$$
whence $$AB + AC + BC + C^2 = 1$$

But each statement contains one mistake, so $AC = 0$, $BC = 0$. Obviously $AB = 0$, hence we are left with $C^2 = 1$, i.e., $C = 1$, which tells us that the name Dibble was correct.

This example, of course, was trivial and was detailed only for clarification of the principles involved. Now we apply those principles to the solution of a slightly more complex problem.

"There's a Jack Brent got married down in Dallas," remarked Sam, looking up from his paper. "That must be Joe's son. Same name and he's twenty-one."

Gwen shook her head. "It's quite a while, and you've forgotten, dear," she told her husband. "His son is Jim, and he'd now be eighteen."

Ann had never met the Brents, but she'd heard plenty about them. "His name certainly wasn't Jack," she informed her mother. "Anyway he's at least twenty-five by now."

Of course all three were wrong one way or another, but each had made one correct statement about either the age or the name.

To find the true first name and age of Joe's son we can conveniently adopt the code:

$$\text{Jack} = a \qquad \text{Age } 18 = d$$
$$\text{Jim} = b \qquad \text{Age } 21 = e$$
$$\text{Not Jack} = c \qquad \text{Age } 25 = f$$

Then Sam said *"ae,"* Gwen said *"bd,"* and Ann said *"cf."* Each made one true and one false statement regarding the name and age, so we have:

$$ae = bd = cf = 0$$

and $\qquad a + e = 1, \qquad b + d = 1, \qquad c + f = 1$

also we must have $ab = ac = de = df = ef = 0$.

Then $\quad (a + e)(b + d) = 1,\quad$ whence $ab + ad + be + de = 1,$

so, dropping the zero-value terms we have: $ad + be = 1$.

Hence, $\qquad\qquad (c + f)(ad + be) = 1$

so $\qquad\qquad acd + bce + adf + bef = 1$

Again dropping zero-value terms we are left with $bce = 1$, which implies $b = 1, c = 1, e = 1$.

So Joe's son was Jim, 21 years old.

Many inferential problems cannot be solved conveniently by the use of Boolean Algebra. These principles, however, can be used advantageously for solving many such problems that may seem quite complex at first. Amongst these are problems that involve the idea of "namesake."

For example, we might have Mr Alton, Mr Bude, Mr Cobb, and Mr Dill; living, not necessarily respectively, in Alton, Bude, Cobb, and Dill.

Complications may seem to arise when, for example, we are told that "the namesake of Mr Bude's home town lives in Dill." But such an item can be dealt with quite simply.

We denote the concept "Mr Alton lives in Cobb," by A_c: similarly, "Mr Bude lives in Alton" by B_a, etc., using the capital letters A, B, C, D for the men and the small letters for the towns. Then, for some one town, at present unspecified but denoted by x, we could write $B_x X_d = 1$. In other words, there is some town x for which it is true that "Mr Bude lives in x, and Mr X lives in Dill."

The method in such problems, then, is to introduce unknowns as they are needed, and then sum over the total possible range of letters (in this case four, but capitals and small letters). In the example here three of the four terms $B_x X_d$ will have zero value, the remaining term therefore having the value 1. And, in the working many of the individual terms can be neglected at sight because they vanish by obvious exclusion relations.

A complete problem will make the idea clear.

Mr Alton, Mr Bude, Mr Cobb, and Mr Dill live in Alton, Bude, Cobb, and Dill, none of the four having the same name as his home town. Dill is not the home town of Mr Alton. The home town of Mr Bude has the name of the man whose home town has the name of the man who lives in Cobb. Where does Mr Dill live?

Using the capital letters A, B, C, D for the men's names, and the small suffixes for the towns, we have:

$A_a = 0$, and similarly for all four letters......................... (1)
$A_d = 0$.. (2)
and there is an X and also a Y such that $B_x X_y Y_c = 1$.......... (3)

We now sum equation (3) for all X and Y, immediately excluding terms which vanish due to equations (1):

$$B_aA_bB_c + B_aA_dD_c + B_cC_aA_c + B_cC_bB_c + B_dD_aA_c + B_dD_bB_c = 1$$

Obviously, $B_aA_bB_c$, $B_cC_aA_c$, $B_cC_dD_c$, $B_dD_bB_c$ all $= 0$, and, from equation (2) $B_aA_dD_c = 0$.

So there remains only $B_cC_bB_c + B_dD_aA_c = 1$, but $B_cC_bB_c$ implies A_a or A_d, *hence* $B_cC_bB_c = 0$. So we are left with $B_dD_aA_c = 1$, from which we conclude that Mr Alton lives in Cobb, Mr Bude in Dill, Mr Dill in Alton, and Mr Cobb in Bude.

Chapter 6

DIOPHANTOS AND ALL THAT

It all started centuries before Diophantos, but he was the first mathematician to make an extensive study of the types of problems and equations that we associate with his name.

Diophantos, the most famous Greek mathematician of his day, lived during the 3rd century: little is known of his life, but we do know that he resided at least for some years in Alexandria about 250 A.D. Of the many books and treatises that he left as a heritage for future generations of mathematicians, some have been lost. Some have been preserved, however, and these include six or seven books in his *Arithmetics* series: all dealing with the properties of rational or of integral numbers.

A typical problem of the type studied by Diophantos would be:

Find two integers such that when one forms their product and adds the square of either integer to it, the result will be a square.

The detailed general solution of this problem will be outlined later in this chapter. It is quoted here merely as an example of a problem that entails solution of what we term a Diophantine equation.

All the problems propounded by Diophantos are not likely to have been originated by him. It seems certain that some had been gleaned from much earlier sources, even from the Babylonians of 1000 and more years before the time of Diophantos. But nothing much is known about the evolution of such problems before his time.

The original problems of Diophantos in most cases called for what we now now describe as "particular solutions," one number or set of numbers satisfying each problem. And the methods he used for solving those problems, and the corresponding equations, were centuries in advance of the general level of mathematical knowledge of those days. Those same methods paved the ways to the evolving of what we now know as general solutions which, as will be seen later, cover all particular solutions of a problem or equation.

So now we can consider the handling of such equations, starting with the simplest. It must be emphasized, however, that we shall be dealing almost exclusively with integral solutions or with rational solutions in which we seek fractional numbers with integral numerators and denominators.

Consider the equation associated with a right-angled triangle, the Pythagorean equation:

$$X^2 + Y^2 = Z^2$$

where X, Y, and Z must be integers (i.e., whole numbers).

There are an infinite number of integral solutions to this equation. For example, $(X,Y; Z) = (3,4; 5)$, or $(5,12; 13)$, or $(8,15; 17)$, or $(6,8; 10)$ are a few particular solutions. But we have to find expressions for X, Y, and Z that will cover all possible integral solutions.

Say $X = kx$, $Y = ky$, $Z = kz$, where x, y, z are integers and k is the greatest integer that is a common factor of X, Y, and Z.

Then, $x^2 + y^2 = z^2$, and x, y, z have no common factor other than unity.

Now, say $z + x = m$, and $z - x = n$, m and n being integers.

Then, $z = \dfrac{m+n}{2}$, $x = \dfrac{m-n}{2}$, and $y^2 = mn$.

To satisfy $y^2 = mn$, m and n being integers, we can say

$$m = rp^2, \quad n = rq^2, \text{ where } p, q, r \text{ are integers.}$$

Substituting these values for m and n, we have

$$z = \frac{r(p^2 + q^2)}{2}, \qquad x = \frac{r(p^2 - q^2)}{2}, \qquad y = pqr$$

Now, in the equation $x^2 + y^2 = z^2$, we can multiply each of x, y, and z by the same number without affecting its validity. Test this in a particular case, if in doubt: say with $(x,y; z) = (3,4; 5)$.

So we multiply each of our derived expressions for x, y, and z by 2, to get:

$$x = r(p^2 - q^2), \qquad y = 2pqr, \qquad z = r(p^2 + q^2)$$

But, by our original definition, unity is the greatest integral common factor of x, y, and z. Hence, in these three expressions we must have $r = 1$, resulting in:

$$x = p^2 - q^2, \qquad y = 2pq, \qquad z = p^2 + q^2,$$

and thence $X = (p^2 - q^2)k$, $Y = 2pqk$, $Z = (p^2 + q^2)k$, which is the required general solution. By assigning any integral values whatsoever to p, q, and k, we will derive an integral solution to the original equation. And any one of the infinity of particular solutions can be obtained by choosing suitable values for p, q, and k.

In deriving this general solution we stipulated k as the greatest integral common factor. Without formal proof, however, it will be obvious that in the final expressions for X, Y, and Z it may be possible in some cases for k to be a fractional number: for example, if p and q are both odd integers we could have $k = 1/2$, or $3/2$, or $5/2$ etc.

We can now undertake the general solution of that simple Diophantine problem that was quoted earlier: to find two integers such that when one forms their product and adds the square of either integer to it, the result will be a square.

The problem can be expressed as

$$\left. \begin{array}{l} x^2 + xy = A^2 \\ y^2 + xy = B^2 \end{array} \right\} \text{where } x, y, A, B \text{ are integers.}$$

Then, $x(x + y) = A^2$, and $y(x + y) = B^2$.

A little thought will show that we must have

$$x + y = u^2 x = v^2 y, \text{ where } u \text{ and } v \text{ are integers.}$$

Then
$$\frac{x}{y} = \frac{v^2}{u^2},$$

which is satisfied by $x = v^2 k$, $y = u^2 k$, k being some integer, and thence $x + y = (u^2 + v^2)k$.

Substituting in the original equations, we get

$$v^2 k^2 (u^2 + v^2) = A^2, \qquad \text{and} \qquad u^2 k^2 (u^2 + v^2) = B^2,$$

so $u^2 + v^2$ must be a square, say $u^2 + v^2 = w^2$.

The general solution of this last equation has been found already, giving us:

$$u = (a^2 - b^2)t, \qquad v = 2abt, \qquad w = (a^2 + b^2)t.$$

So we can now substitute these values for u and v in the stipulated $x = v^2k, y = u^2k$,

hence $\qquad\qquad x = 4a^2b^2t^2k, \qquad y = (a^2 - b^2)^2t^2k.$

But t and k are common factors which can be chosen at will, so there can be no loss of generality or validity if we say $t^2k = s$, where s can be any common factor.

The final general solution then becomes:

$$x = 4a^2b^2s, \qquad y = (a^2 - b^2)^2s.$$

It is always only too easy to make a mistake when solving such an equation, and for that reason it is wise to check any solution: in more complex cases, it is often wise to check the intermediate results at different stages of the working.

Here our solution can be checked by direct substitution of the derived expressions for x and y, which you may care to do as a little exercise in algebraical manipulation. But even a check for one particular set of values can be useful, even though it does not provide conclusive confirmation as to the correctness of the solution for all values.

Say $a = 5, b = 2, s = 3$. Then $x = 1200, y = 1323$, and we have

$$1200^2 + 1200 \cdot 1323 = 1200 \cdot 2523 = 1740^2$$
and $\qquad 1323^2 + 1200 \cdot 1323 = 1323 \cdot 2523 = 1827^2$

In many problems involving Diophantine equations we have to find the smallest integers satisfying the equation, so let us see how that would be done in this case.

If $b = 0$, we would have $x = 0$, and obviously a zero value is not acceptable as a reasonable result. So, for our "smallest solution," we would take $a = 2$, and $b = 1$. This would lead to $x = 16s$, $y = 9s$. And, because 16 and 9 have no common factor other than unity, our "smallest values" come with $s = 1$, giving $x = 16, y = 9$.

Considering the original conditions, however, it will be seen that x and y must be interchangeable. So we would expect there to be values for a, b, and s such that this interchange would result.

Putting a and b equal to two odd numbers must make both $a + b$ and $a - b$ multiples of 2, so that $a^2 - b^2$ will then be a multiple of 4. The "co-ordinating constant," s in our solution, can be a fraction and in this case it could obviously be $\frac{1}{4}$.

So, with $a = 3, b = 1, s = \frac{1}{4}$, we have $x = 9, y = 16$.

Hence we have established the smallest solution in this problem as $(x,y) = (16,9)$, which is a short way of writing that $x = 16$, $y = 9$, with x and y interchangeable.

We now move on to an extension of that first simple Pythagorean equation. The solution will be stated, but without any attempt to show how it can be derived.

Say we have $X^2 + eY^2 + fZ^2 = W^2$, where e and f are any integral coefficients, positive or negative.

The simplest general solution of this is:

$$
\left.
\begin{aligned}
X &= \pm(a^2 - eb^2 - fc^2)k \\
Y &= 2abk \\
Z &= 2ack \\
W &= \pm(a^2 + eb^2 + fc^2)k
\end{aligned}
\right\}
\quad
\begin{aligned}
&\text{where } a,\, b,\, c \text{ are integers, and} \\
&\quad k \text{ is any common factor}
\end{aligned}
$$

In this solution note that X^2 remains positive whether X be positive or negative, and similarly Y^2 and Z^2 and W^2 are not affected by change of sign. In most practical cases we would be concerned only with positive values of X, Y, Z, and W: but this would entail use of the minus sign in the expression for X, if a, b, and c were such that $a^2 < eb^2 + fc^2$.

For example, say we have $X^2 + 2Y^2 - 3Z^2 = W^2$. Then the simplest general solution is:

$$
\begin{aligned}
X &= \pm(a^2 - 2b^2 + 3c^2)k \\
Y &= \pm 2abk \\
Z &= \pm 2ack \\
W &= \pm(a^2 + 2b^2 - 3c^2)k
\end{aligned}
$$

and, setting $a = 2$, $b = 2$, $c = 1$, $k = 1$, we have one particular solution in positive integers as

$$X = 1, \qquad Y = 8, \qquad Z = 4, \qquad W = 9$$

This equation and its general solution, as given, can be most useful in the solving of many more complex diophantine equations. We can now look at a few examples of such use.

Say we wish to solve $x^2 + 2y^2 = z^2$ in positive integers. Comparing this with $X^2 + eY^2 + fZ^2 = W^2$, it will be seen that we have $e = 2$, $f = 0$, so that the required solution becomes

$$
\begin{aligned}
x &= (a^2 - 2b^2)k \\
y &= 2abk \\
z &= (a^2 + 2b^2)k
\end{aligned}
$$

A rather more difficult example would be the solution of a quite reasonable little teaser:

"Bill, Bert, and Bob are brothers, there being no twins in their family. Ignoring odd months in their ages, the sum of the squares of Bill's and Bert's ages equals five times the square of Bob's age. How young can Bob be?"

We say the ages in years are X, Y, and Z, where Z is Bob's age, and so we have $X^2 + Y^2 = 5Z^2$. Bearing in mind that there are no twins, this can be solved at sight by $X = 11$, $Y = 2$, $Z = 5$. But let us derive the theoretical solution which would obviate much very tedious trial and error if larger numbers were involved.

The equation can be written as $X^2 + Y^2 - (2Z)^2 = Z^2$.

Omitting, but not forgetting, the "co-ordinating constant" in our previous general solution, this equation is satisfied by

$$X = a^2 - b^2 + c^2, \qquad Y = 2ab,$$
$$Z = ac = a^2 + b^2 - c^2$$

Then, $\qquad c^2 + ac - a^2 = b^2$

so $\qquad 4c^2 + 4ac - 4a^2 = 4b^2$

i.e., $\qquad (2c + a)^2 - 5a^2 = (2b)^2$, which is satisfied by:

$$2c + a = m^2 + 5n^2, \qquad 2b = m^2 - 5n^2,$$
$$a = 2mn, \text{ omitting a co-ordinating constant.}$$

So we have: $2c = m^2 - 2mn + 5n^2$, $2a = 4mn$, $2b = m^2 - 5n^2$. In such a solution we can multiply or divide each expression by any common multiplier or divisor without affecting its validity, so we re-write as:

$$a = 4mn, \qquad b = m^2 - 5n^2, \qquad c = m^2 - 2mn + 5n^2$$

Substituting these values for a, b, and c, and simplifying the resulting expressions, we get:

$$X = 4mn(10mn - m^2 - 5n^2)$$
$$Y = 4mn(2m^2 - 10n^2)$$
$$Z = 4mn(m^2 - 2mn + 5n^2)$$

Initially we omitted a co-ordinating constant. We now see $4mn$ as a common factor in the three expressions. Without in any way affecting the validity of the solution, that common factor $4mn$ may be removed and in its place we can insert a completely independent co-ordinating constant "k." And, since any or all of the three expressions may be multiplied by "-1" without affecting the validity of the solution, it will be as well to apply this device in the expression for X.   Our final general solution then becomes:

$$X = \pm(m^2 - 10mn + 5n^2)k$$
$$Y = \pm 2(m^2 - 5n^2)k$$
$$Z = \pm(m^2 - 2mn + 5n^2)k$$

Setting $m = 2$, $n = 1$, $k = 1$, we have positive values: $X = 11$, $Y = 2$, $Z = 5$, so 5 years is the youngest age for Bob.

Another type of diophantine equation provides a good example of a very different treatment.   Say we have:

$$\left. \begin{array}{l} x^2 + y = A^2 \\ y^2 + x = B^2 \end{array} \right\}$$ where x, y, A, B must be rational non-zero numbers, not necessarily integers.

We can put $A = x - m$, $B = y - n$, where m and n must be rational numbers, not necessarily integers.

Then, $x^2 + y = x^2 - 2mx + m^2$,   and $y^2 + x = y^2 - 2ny + n^2$, so $y = m^2 - 2mx$,   and $x = n^2 - 2ny$,

combining these, $\left. \begin{array}{l} 2mx + y = m^2 \\ x + 2ny = n^2 \end{array} \right\}$

i.e., $\left. \begin{array}{l} 4mnx + 2ny = 2m^2n \\ x + 2ny = n^2 \end{array} \right\}$

whence, $x = \dfrac{n(2m^2 - n)}{4mn - 1}$ and similarly $y = \dfrac{m(2n^2 - m)}{4mn - 1}$,

which is the desired general solution for x and y.

This solution raises some points that are worthy of our consideration here.

Firstly, it should be noted that the only non-zero integral solution for x and y is given by putting $m = -1$, $n = -1$, leading to $x = -1$, $y = -1$, but making $A = B = $ zero.   The formal proof is rather outside the scope of these pages, but it can be proven that for all other non-zero solutions both x and y must be fractions.   For example, $x = y = \frac{1}{3}$, is an acceptable solution.

If we require both x and y to be positive, then m and n must be chosen so that $2m^2 > n$, and also $2n^2 > m$.   These conditions can be combined as:

$$2n^2 > m > \sqrt{\frac{n}{2}}$$

For any selected value of n, it then becomes a simple matter to ascertain the permissible range of values for m. For example, with $n = 2$, we have $8 > m > 1$, and positive values for x and y will result from any chosen m within that range. With $m = 3$, $n = 2$, we have:

$$x = \tfrac{32}{23}, \qquad y = \tfrac{15}{23}, \qquad A = \tfrac{37}{23}, \qquad B = \tfrac{31}{23}$$

No introduction to diophantine equations, however brief, would be complete without mention of the Pell equation. This is a particular type of diophantine equation, so named for the mathematician who first singled it out for special attention. In its simplest form the Pell equation is

$$x^2 - 2y^2 = 1.$$

This has an infinite number of solutions in integers. The successive pairs of values can be tabulated as:

$$x = 1 \quad 3 \quad 17 \quad 99 \quad 577 \quad \text{etc.}$$
$$y = 0 \quad 2 \quad 12 \quad 70 \quad 408 \quad \text{etc.}$$

Notice the connection between the respective values of x and y in successive pairs in this series of solutions, starting with the solution next after the first non-zero solution:

$$17 = 6 \cdot 3 - 1, \quad 99 = 6 \cdot 17 - 3, \quad 577 = 6 \cdot 99 - 17, \quad \text{etc.}$$
$$12 = 6 \cdot 2 - 0, \quad 70 = 6 \cdot 12 - 2, \quad 408 = 6 \cdot 70 - 12, \quad \text{etc.}$$

We should also note that $x = 3$ is the value for x in the first non-zero solution, and that $6 = 3 \times 2$.

Next we take the equation $x^2 - 3y^2 = 1$, the successive pairs of integral solutions being:

$$x = 1 \quad 2 \quad 7 \quad 26 \quad 97 \quad \text{etc.}$$
$$y = 0 \quad 1 \quad 4 \quad 15 \quad 56 \quad \text{etc.}$$

Here we have:

$$7 = 4 \cdot 2 - 1, \quad 26 = 4 \cdot 7 - 2, \quad 97 = 4 \cdot 26 - 7, \quad \text{etc.}$$
$$4 = 4 \cdot 1 - 0, \quad 15 = 4 \cdot 4 - 1, \quad 56 = 4 \cdot 15 - 4, \quad \text{etc.}$$

In this case, $x = 2$ is the value for x in the first non-zero solution, and $4 = 2 \times 2$.

Continuing this process, the next case would appear to be $x^2 -$

$4y^2 = 1$. But this is $x^2 - (2y)^2 = 1$, which can have no non-zero solution at all. Similarly, we would have to ignore such cases as $x^2 - 9y^2 = 1$, $x^2 - 16y^2 = 1$, etc.

So the next case to consider will be $x^2 - 5y^2 = 1$, which has successive pairs of solutions:

$$x = 1 \quad 9 \quad 161 \quad 2489 \quad \text{etc.}$$
$$y = 0 \quad 4 \quad 72 \quad 1292 \quad \text{etc.}$$

Here we have:

$$161 = 18 \cdot 9 - 1, \quad 2489 = 18 \cdot 161 - 9, \quad \text{etc.}$$
$$72 = 18 \cdot 4 - 0, \quad 1292 = 18 \cdot 72 - 4, \quad \text{etc.}$$

In this case, $x = 9$ is the value for x in the first non-zero solution, and $18 = 9 \times 2$.

To continue the process would be mere waste of time. It can be proven, in fact, that in every case of the general Pell equation $x^2 - Ny^2 = 1$ (except where N is a square), the successive pairs of integral solutions, after the first non-zero pair, obey the absolute rule:

$$x_n = 2ax_{n-1} - x_{n-2} \left.\right\} \text{ where } x = a \text{ is the value of } x$$
$$y_n = 2ay_{n-1} - y_{n-2} \left.\right\} \text{ in the first non-zero solution.}$$

This is an important and most useful rule, which makes it possible to write down the successive pairs of solutions for such an equation virtually at sight, once the first non-zero solution has been found. There is a way to calculate the first non-zero solution, but in most cases it is quicker and more practical to find this by trial: trying successive values for y until one reaches the value which will yield an integral value for x.

Most text-books, incidentally, give a far more complex and quite impracticable method for obtaining the required solutions. The method outlined here is no less sound, and is almost always preferable.

We have considered the general Pell equation in the form $x^2 - Ny^2 = +1$, in which there are an infinite number of integral solutions for each and every possible value of N (except if N be a square).

We can now extend our survey to the even more general form $x^2 - Ny^2 = e$, where e is some positive or negative integer other than $+1$, and where N is not a square. This will include, as a particular case, $x^2 - Ny^2 = -1$.

Here we cannot by any means claim that there will be integral solutions for all values of N and e. In fact, for any particular value of e, integral solutions will exist only for certain values of N, and vice versa.

For example, if $e = -1$, so that $x^2 - Ny^2 = -1$, there will be integral solutions only with $N = 2$, or 5, or 10, or 13, or 17, etc. If $N = 3$, there will be integral solutions only where $e = +1$, or -2, or -3, or $+4$, or $+6$, or -8, etc. The theoretical problem of ascertaining such *compatible* values for N and e is very complex, but in practice the necessary confirmation as to whether or not integral solutions exist for any given N and e can be found by actual testing over a reasonably limited range of successive values for y.

Let us assume then that we have ascertained that there is indeed an integral solution for $x^2 - Ny^2 = e$, for some particular values of N and e. And let us assume that this be the integral solution that involves the smallest values of x and y, as $x = a$, $y = b$. Also, let $x = m$, $y = n$, be *any* integral solution of the equation $x^2 - Ny^2 = +1$, the N having the same given value as in the original equation. Then, $x^2 - Ny^2 = e = a^2 - Nb^2$, and $m^2 - Nn^2 = 1$, so

$$
\begin{aligned}
x^2 - Ny^2 &= (a^2 - Nb^2)(m^2 - Nn^2) \\
&= a^2m^2 + N^2b^2n^2 - Na^2n^2 - Nb^2m^2 \\
&= a^2m^2 \pm 2Nabmn + N^2b^2n^2 - Na^2n^2 \mp 2Nabmn - Nb^2m^2 \\
&= (am \pm Nbn)^2 - N(an \pm bm)^2
\end{aligned}
$$

We can now say, $x = am \pm Nbn$, and $y = an \pm bm$, noting that the sign (i.e., $+$ or $-$) must be the same in each. Substituting for m and n any pair of values that satisfies the equation $m^2 - Nn^2 = 1$, we will obtain solutions to the original equation $x^2 - Ny^2 = e$.

An example will make this clear. Say we have $x^2 - 3y^2 = -11$, the smallest integral solution of which is $x = 1$, $y = 2$. Then, the required solution is $x = m \pm 6n$, $y = n \pm 2m$. But, remember that $(-x)^2 = (+x)^2$, and $(-y)^2 = (+y)^2$. So we can write this solution in a more "tidy" way as:

$$
\begin{aligned}
x &= \pm(m \pm 6n) \\
y &= \pm(2m \pm n)
\end{aligned}
\left. \right\}
\begin{array}{l}
\text{where } m \text{ and } n \text{ are integers} \\
\text{such that } m^2 - 3n^2 = 1
\end{array}
$$

We can tabulate successive values of m and n, as already explained, and so can write down corresponding x and y:

$m = 1$	2	2	7	7	26	26	etc.
$n = 0$	1	1	4	4	15	15	etc.
$x = 1$	4	8	17	31	64	116	etc.
$y = 2$	3	5	10	18	37	67	etc.

By continuing this tabulation, for the further successive m and n values, we could derive all possible integral solutions for x and y up to any limit desired. So, in this particular case where our equation is $x^2 - 3y^2 = -11$, this procedure can be said to give all integral solutions.

It is possible, however, to have an equation of this type for which the procedure as outlined will not give all the integral solutions. For such an equation some modification in the procedure is required. This can best be illustrated with a practical example.

Say we wish to find all integral solutions of $x^2 - 2y^2 = 119$, subject to x being less than 200.

By quick trial we find that the smallest solution is $x = 11$, $y = 1$. Hence, by the procedure already explained, we can write the general form for other integral solutions as:

$$x = \pm(11m \pm 2n), \qquad y = \pm(m \pm 11n), \qquad \text{where } m^2 - 2n^2 = 1$$

Using the successive values of m and n, we derive values:

$x = 11$	29	37	163
$y = 1$	19	25	115

In fact, there are other solutions within the limit for x. Whether or not such other solutions exist should always be checked, and the method of checking this is not very difficult.

We obtained the tabulated values for x and y by using the smallest integral solution for the equation, $x = 11$, $y = 1$. If there are other solutions, quite independent of those we derived from that smallest integral solution, then the smallest of such *other* solutions must give values for x and y between the **smallest and the next smallest** solutions as shown in our first tabulation: in this case, between $x = 11$, $y = 1$, and $x = 29$, $y = 19$.

So, to carry out the necessary check, we test in the original equation for successive values of y from $y = 2$ to $y = 18$. This may seem rather laborious, but it can be done quickly and without undue figuring.

Checking in this way, with our equation $x^2 - 2y^2 = 119$, we soon find a new solution $x = 13, y = 5$.

Now we can tabulate a second series of values for x and y, based on $x = 13, y = 5$. The values for x and y follow the same principles as those in the previous series, as:

$$x = \pm(13m \pm 10n), \qquad y = \pm(5m \pm 13n), \qquad \text{where } m^2 - 2n^2 = 1,$$

and we derive:

$x =$	13	19	59	101
$y =$	5	11	41	71

The process can be repeated, just in case there is yet a third series (or *family*) of solutions independent of the two already derived. If there be such another series, then the smallest solution in it must give a value for y between the **smallest and next smallest** values just found in the second series. So we would test successive values of y from $y = 6$ to $y = 10$: if this revealed a solution, then the procedure would have to be repeated with a third tabulation based on that solution.

In fact, there is no third "family" of solutions in the case of $x^2 - 2y^2 = 119$. So, to make things clearer, we can now tabulate our complete results showing the two distinct "families" as follows:

$x =$	11			29	37			163
$y =$	1			19	25			115
$x =$		13	19			59	101	
$y =$		5	11			41	71	

It may be noted that 119 is the product of 7 and 17, both prime numbers. The example that has been worked through exemplifies a useful rule, which would have told us that there could be two but not more than two distinct families of solutions for the equation $x^2 - 2y^2 = 119$. This may be stated in simple terms as:

In the equation $x^2 - Ny^2 = e$, where N is not a square, and where e and N have no common factor, if e is the product of m odd prime numbers (no factor of e being a square or higher power), then there will be at most 2^{m-1} families of integral solutions or else no integral solutions at all. *Note:* If e is a multiple of N, then the equation is not in its most primitive form. This can be seen by assuming $e = fN$, in which case x must also be a multiple of N and we can say $x = zN$. Substituting these values, we derive a new equation $y^2 - Nz^2 = -f$. Since f has fewer prime factors than e, the stated rule would apply to the factors of f rather than to those of e. Consideration of cases where e and N have a common factor, although e is not a multiple of N, is more complex and so outside our scope here.

For example, the equation $x^2 - 11y^2 = 9805$ has four distinct families of solutions, 9805 being the product of 5, 37, and 53. The reader may feel tempted to find all solutions for positive values less than $y = 250$.

Where e is even, and or with squares or higher powers of primes amongst its factors, things become rather more complex and it would be difficult to discuss such cases here. But the general principles would still apply so far as the finding of the relevant families of solutions is concerned.

The practical aspects in connection with solving these Pell equations have been considered in some detail. Such equations arise in many *recreational* problems, rarely involving *difficult* numbers, and this brief survey has been intended to provide some understanding of the methods to be used in their solution. At the same time it must be emphasized that this survey has been by no means comprehensive: many large books have been written on the Pell equation alone!

In general, this chapter has only partially covered a very small part of the whole field of diophantine equatiohs. It is hoped, however, that it will have given a small insight into the elements of what is involved when solving such equations. Further examples, although possibly not explained in detail, will be found later in this volume in the solutions to some of the teasers.

Chapter 7

POTPOURRI

The dictionary defines a potpourri as a collection of unrelated items, and that is precisely what this chapter comprises. Some of the items might well have been included in appropriate chapters elsewhere in this book, but we feel that those chosen for inclusion here are of sufficient individual interest to justify their being picked for special mention.

Geometric Dissections

These provide one of the most popular pastimes in recreational mathematics. This may be because little knowledge of mathematics is required: intuition, trial-and-error, and patience usually prove far more successful than any attempt to apply mathematical theory when solving a dissection problem.

In the most common form, such a problem requires the conversion of one geometrical figure into another by cutting the former into the least number of pieces which, when rearranged, will form the new figure. It has been proved that any polygon, however irregular, can be cut into a finite number of pieces which will reform into any other specified polygon of the same area.

Later in this chapter other types of dissections will be discussed. But here are a few typical geometric dissections, showing how a Latin cross may be converted into various other geometrical figures. They are the work of Harry Lindgren, and are believed to be the solutions with the least number of pieces in each case: no proof, however, has been found confirming this statement.

The three figures, 7-1, 7-2, and 7-3 illustrate two of Lindgren's methods for working out these dissections.

A Latin Cross can be converted into a Golden Rectangle [where the length to width ratio is $(\sqrt{5} + 1)$ to 2] by utilizing a "strip" method. The Latin Cross is itself dissected so that the parts can be arranged to form an endless strip with parallel sides. The Golden Rectangle need not be dissected, for a series of them end-to-end will

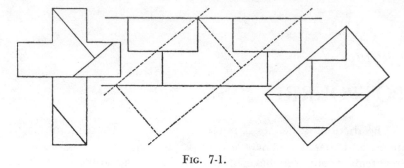

FIG. 7-1.

form the required strip. One strip is placed on top of the other (it is advisable to draw the strips on transparent, or tracing, paper) and turned in various ways. A satisfactory result comes about only when the edges of one of the strips pass through congruent points of the other. Repeated attempts are tried until one is satisfied that he has achieved a dissection with the least number of pieces.

Figure 7-2, showing the dissection of a Latin Cross to a hexagon, also utilizes the strip technique.

A more sophisticated method is shown in the Latin Cross-dodecagon dissection. The dodecagon can be dissected as shown in Figure 7-3A to form a pattern that can be repeated in a plane—commonly referred to as a *tesselation* pattern. The Latin Cross can also form a tesselation pattern as shown. The necessity here is that both tesselation patterns must have a congruent set of points—which are indicated by the circled dots. Figure 7-3B shows the two tessela-tions superimposed with the resulting dissection.

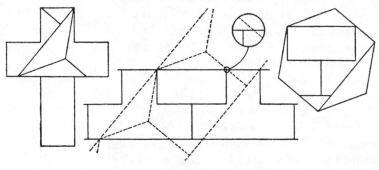

FIG. 7-2.

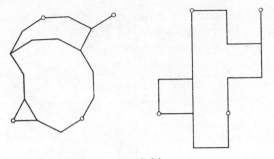

FIG. 7-3A.

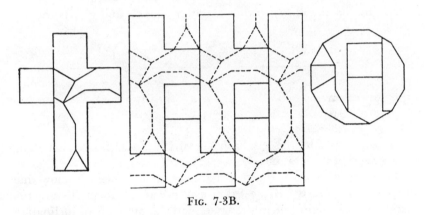

FIG. 7-3B.

The Ham Sandwich Theorem

It is obvious that any line segment has only one point which bisects it.

It may not be so obvious that there is only one line which mutually bisects two line segments in space—the line segments not being in line with each other. The proof of this is simple: since each line segment has only one point bisecting it, a line connecting those two midpoints must bisect both.

Any three points in space lie upon the arc of some circle (a straight line segment may be considered to be the arc of a circle of infinite radius), so a similar argument shows that there is an arc of a circle which will mutually bisect any three line segments in space—again provided no two of the three are in line with each other.

What about the following statement? "Any two areas in a plane can be mutually bisected by a straight line."

In Figure 7-4 we have two rectangles, and a line passing through

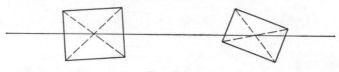

FIG. 7-4.

their center points. This line obviously bisects both rectangles.

Figure 7-5 shows two quite irregular shapes, and a hypothetical

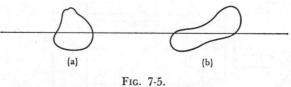

(a)   (b)

FIG. 7-5.

straight line which bisects both areas. We now prove that there can be such a straight line.

Considering each of the two shapes separately, we can show that each can be bisected by an infinite number of straight lines. Imagine some straight line in the same plane, at an angle of inclination θ to an arbitrary reference line XY. See Figure 7-6. Sweeping

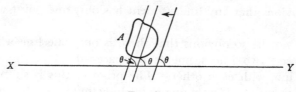

FIG. 7-6.

this line, at that angle θ to XY, through the shape A, clearly there will be one position where it will bisect the area of that shape. Since there are an infinite number of lines that bisect the area of shape A, there will always be one such position corresponding to whatever value the angle θ may have.

Precisely similar considerations can be applied to shape B. So one of the infinitude of lines that will bisect the area A must co-incide with one of the infinitude of lines that will bisect area B. This will be the line that mutually bisects both areas.

Extending these ideas into three dimensions it can be proved that any three solids in space may be mutually bisected by some single plane. The proof is rather beyond the scope of this book, but the theorem does have an amusing practical application. No matter how one arranges a single slice of ham between two slices of bread, one being brown bread and the other white, there is always at least one way in which the sandwich can be sliced through by one clean cut that will divide the ham, and also each type of bread, into equal portions! Sad to say, the theorem gives no hint as to how that ideal cut may be made.

Whilst on the subject of food, here are two little problems connected with equitable division. The answers will be found on page 129.

(1) How, with only a knife available, can two people divide a piece of cake so that each will be completely satisfied that he has received his fair half share?

(2) Ahmed, Kemal, and Ali had been given a sack of raisins to share among themselves. How could they do this fairly, and to the satisfaction of each, without any form of measuring or weighing equipment?

Cube Formation

A cube formation problem was proposed in *Recreational Mathematics Magazine**: "Cut a 1 by 3 rectangle of paper into two identical pieces that will form a perfect cube when folded and joined (without any overlapping)."

This was thought to have only two solutions, as shown in Figure 7-7, where the dotted lines indicate folds.

Fig. 7-7.

* RMM No. 7, Feb. 62, p. 24.

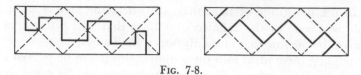

FIG. 7-8.

Two more distinct straight-cut solutions were discovered, however, by C. L. Baker, as shown in Figure 7-8.

Furthermore Baker discovered an infinite number of solutions to the problem! In Figure 7-9 we start by drawing *any* curve connecting points *A* and *B*. This curve is then turned through 90° about point *B*, to connect *B* and *C*. It is then turned through 180° about point *C*, to connect *C* and *D*. A further rotation of the whole curve, *A* to *D*, about point *D* connects *D* to *E*. Cutting carefully along this complete curve, *A* to *E*, and then folding along the previously marked dotted lines, the two pieces will form a cube when joined together.

Restrictions in this ingenious procedure are that one must avoid a curve which would "cross" and so lead to more than two pieces, and of course the curves must be contained within the boundaries of the paper strip.

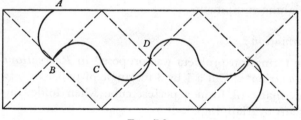

FIG. 7-9.

A Point in Topology

We stretch a short length of rubber string exactly along the imaginary straight line of which it virtually forms a line segment, the string in its stretched state containing all of that line that was within it in its unstretched state.

Can we show that there must be at least one point in the string that remained fixed despite the distortion? In other words, can we prove that at least one point in the string must be in precisely the same position after the stretching as it was in before the stretching?

In Figure 7-10 we represent the two states of the string by the short and the long parallel lines.
For each and every point in the unstretched string there must be a corresponding point in the stretched string: lines are drawn to illustrate this. Then at least one line, drawn to connect corresponding points, must be perpendicular to both the lines that represent states of the string. This line connects the two representations of one point that has remained fixed.

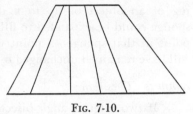

Fig. 7-10.

It may be worth mentioning that no connecting line here could cross any other connecting line. Such a crossing would imply that two points in the string had *exchanged* relative directions, one from the other, along the length of the string: an obvious impossibility in any lengthwise stretching.

Can the same principle be applied to a two-dimensional plane? Say we have a circular rubber disc A (see Figure 7-11), and then stretch it in its plane so that it takes shape B, it being a condition that the whole of A must lie within B. Must there be at least one point that has remained undisturbed by the stretching?

In Figure 7-12 we represent the two states of the disc by the configurations A and B, understood to lie parallel to each other. As in the case of the stretched string, here too there is a point-to-point (i.e., one-to-one) correspondence between the two representations. This is illustrated in Figure 7-12. Clearly, the line connecting

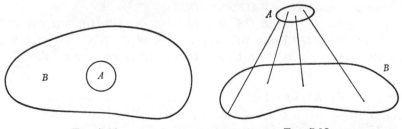

Fig. 7-11.

Fig. 7-12.

some one particular pair of corresponding points must be perpendicular to both configurations. This line connects the two representations of one point that has remained fixed.

Not too surprisingly, this concept applies in three dimensions: in fact it applies theoretically in n-dimensional space to the stretching of an object of up to n dimensions. So, for example, if a sponge could be expanded to fill a room there is certainly one fixed point in that sponge—a point whose precise position in the room will have remained unchanged by the expansion.

Lehmus' Theorem

"If two interior angle-bisectors in a triangle are equal, the triangle must be isosceles."

Lehmus proposed this theorem to Jacob Steiner in 1840, and ever since then it has been the subject of much discussion and many a mathematical paper. Its beauty lies in the simplicity of the statement, and in the difficulty of proving its truth. Some proofs have occupied many pages of print, others only a few lines. The proof we give here, by J. A. H. Hunter, is one of the shortest known and has not been published previously.

In a triangle ABC, the bisectors of the interior angles at A and B are drawn to meet sides BC and AC at Q and R respectively. We must prove that the triangle is isosceles if $AQ = BR$.

Proof

First draw the triangle ABC (Figure 7-13) obviously not isosceles, with $\angle BAC > \angle ABC$, and draw the angle bisectors AQ and BR. Draw PQ and RS parallel to AB, and draw QU parallel to CA. Let BR produced and QP produced meet at T.

To justify this construction, we first prove that if P lies "above" R, then we must have $\angle BAC > \angle ABC$.

$QU = PA = PQ$ (since QA is a bisector of $\angle BAC$), but $PQ < TQ$, and $TQ = QB$ (since BR is a bisector of $\angle ABC$). So $\angle QUB > QBU$, hence $\angle BAC > \angle ABC$. This justifies drawing P "above" R.

Now, $PQ < RS$, so $PA < BS$ .... (1)

Also, $AQ = BR$ is our basic assumption, and we have $\angle SBR < PAQ$, so on that basic assumption $BS < PA$ .... (2)

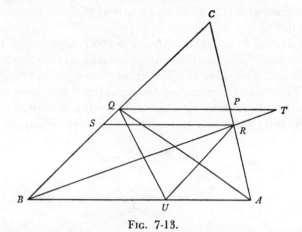

FIG. 7-13.

Inequalities (1) and (2) are contradictory, hence the triangle ABC cannot be non-isosceles. So it must be isosceles.

Picking Particular Permutations

There are 720 ways in which 6 different things can be arranged in order, 5040 ways for 7 different things. More precisely, we say there 720 permutations of 6 different things, 5040 permutations of 7 different things. And, in general, the number of permutations of n different things is given by "factorial n", shown as $n!$, where:

$$n! = n(n - 1)(n - 2) \ldots (2)(1)$$

For example, $6! = 6 \cdot 5 \cdot 4 \cdot 3 \cdot 2 \cdot 1 = 720$.

A high-school student, Dale Kozniuk, has drawn our attention to a very interesting point in connection with such permutations, an idea that comes well within the scope of these pages.

It is a simple matter to evaluate 8! say, and so find that we can arrange the 8 digits—1 through 8—in 40,320 different ways. Given the necessary time and much patience, we could even list those "ways" in numerical order starting with 12345678, 12345687, 12345-768, 12345786, etc.

But by listing those successive "numbers," it would be a long time before we had established the identity of, say, the 20,117th permutation in that ordered tabulation!

Fortunately, any particular permutation can be identified quickly and with comparatively little calculation, without compilation of that long list.

Before discussing this procedure, however, there is one fact that must be clarified in connection with the factorial function. Clearly $3! = 6$, $2! = 2$, and $1! = 1$. But it may not be so obvious that $0! = 1$. This can be proved, but the rigid proof is outside our scope here. Although no formal proof, the following succession of similar operations will suggest that "factorial zero" indeed equals unity.

$$3! = \frac{4!}{4} = \frac{24}{4} = 6$$

$$2! = \frac{3!}{3} = \frac{6}{3} = 2$$

$$1! = \frac{2!}{2} = \frac{2}{2} = 1$$

$$0! = \frac{1!}{1} = \frac{1}{1} = 1$$

Say we wish to find the 2238th permutation in a complete tabulation of the 5040 permutations of the digits 1 through 7, the permutations being arranged in increasing "numerical order" (i.e., starting with 1234567, 1234576, 1234657, etc.).

First we break down that number 2238 into multiples of the factorials of the numbers less than 7 and including 0!.

$$2238 = 3 \cdot 6! + 0 \cdot 5! + 3 \cdot 4! + 0 \cdot 3! + 2 \cdot 2! + 1 \cdot 1! + 1 \cdot 0!$$

In this breakdown we exemplify a vital point. Having taken $3 \cdot 720 + 0 \cdot 120 + 3 \cdot 24$, totalling 2232, there is a remainder of 6. And we know that $3! = 6$, which might suggest that this remainder could be broken down as $1 \cdot 3! + 0 \cdot 2! + 0 \cdot 1! + 0 \cdot 0!$. Wherever, in the process of breaking down, we derive a "remainder" that can be broken down in more than one way, we must adopt the breakdown which "uses" as much as possible the non-zero multiples of the factorials of the smaller number or numbers. Similarly, if in another case, we reached a remainder of 5, we would break it down as $2 \cdot 2! + 0 \cdot 1! + 1 \cdot 0!$ and not as $2 \cdot 2! + 1 \cdot 1! + 0 \cdot 0!$.

Now, referring to the "coefficients" of the several factorials in our breakdown of 2238, we write down successive ordinal numbers, each being 1 greater than the corresponding "coefficient" but the last ordinal number in the series being exactly equivalent to the "coefficient" of 0!:

4th,   1st,   4th,   1st,   3rd,   2nd,   1st

We than proceed to identify the digits in the 2238th permutation by the following method, the operation of which will be obvious:

Remaining digits in ascending order	Ordinal Number	Required Digit
1 2 3 4 5 6 7	4th	4
1 2 3 5 6 7	1st	1
2 3 5 6 7	4th	6
2 3 5 7	1st	2
3 5 7	3rd	7
3 5	2nd	5
3	1st	3

Then, having found the required digits, the 2238th permutation must be 4162753.

This result can be checked quite simply. A complete list of all the permutations, from 1234567 right up to 7654321, would comprise 5040 numbers in increasing order. Hence the 2238th permutation in that list must be identical with the 2803rd number in a reversed tabulation from 7654321 down to 1234567.

$$2803 = 3 \cdot 720 + 5 \cdot 120 + 1 \cdot 24 + 3 \cdot 6 + 0 \cdot 2 + 0 \cdot 1 + 1 \cdot 1$$

So we have 4th, 6th, 2nd, 4th, 1st, 1st, 1st.

Following exactly the same method, but this time with digits 7654321 in descending order, we would evaluate the 2803rd permutation (in our reversed list) as 4162753, which checks with the original evaluation.

A Rather Large Number

There is a popular mathematical game in which one has to form expressions for numbers using certain stipulated numerals with regular arithmetical signs and symbols.

Using one each of the digits 1, 2, 3, and 4 together with only decimal points, minus signs, and brackets, a fantastically large integer can be expressed.

We have $N = 3^{(.2)^{-(.1)^{-4}}} = 3^n$, where n has approximately 6990 digits.

The number of *digits* in N is a *number* with roughly 3000 digits.

In comparison with the value of N, the number of cubic inches in the whole volume of space comprising the observable universe is an almost negligible quantity!

Chapter 8

FUN WITH SHAPES

The search for mathematical proof has led many down the path of sheer frustration. The famous four-color map theorem states that no more than four colors are ever required to color a map on a plane or spherical surface. No one has ever devised a map that requires five colors—but this is no *proof* that only four are necessary. Just one map requiring five colors would settle a question that has troubled mathematicians for centuries; or a firm proof that no more than four colors are necessary.

Some problems, however, look much more formidable than they really are.

Figure 8-1 shows an 8 by 8 array of squares, with one square removed from each of two opposite corners. Can we cover the remaining 62 squares with 31 dominoes? Each domino would cover 2 squares, so on the face of it this might seem possible. But all attempts to do it must fail. And the real problem is to *prove* that this must be so. Dozens, hundreds, and even thousands of abortive attempts to cover the 62 squares with 31 dominoes constitute no firm proof of its impossibility.

If we color the squares, as shown in Figure 8-2, we have a regular checkerboard pattern, and it will be seen that the two missing corner squares would both be white or both black: in the figure

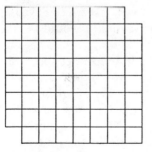

FIG. 8-1.

FIG. 8-2.

77

they would be white. There remain, then, 32 black and only 30 white squares. But a domino must necessarily cover one black and one white square, so 31 dominoes would cover 31 squares of each color. Since we have but 30 white squares, the 31 dominoes cannot cover the 62 squares as required.

There is a somewhat similar problem in connection with 3-dimensional dominoes. Here we have a 3 by 3 by 3 cubical structure containing 27 cubes, and 3-dimensional dominoes each comprising one black and one white cube. We have to remove one of the 27 cubes in the structure, and then replace the 26 remaining cubes by 13 of the 3-dimensional dominoes. The problem lies in determining which of the 27 cubes may be removed, or alternatively which of them cannot be removed so that the required replacement can be made.

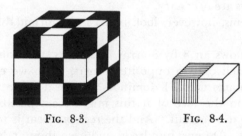

FIG. 8-3.   FIG. 8-4.

The solution of this is really quite simple. We color the alternate cubes in the structure, the center cube remaining white (see Figure 8-3). Then we have 14 black cubes and 13 white. The 13 3-dimensional dominoes comprise 13 black and 13 white cubes, so we must remove a black cube from the structure in order to leave 13 black and 13 white cubes corresponding to the 13 dominoes: it is immaterial which of the black cubes is removed. If we removed a white cube, so leaving 14 black and 12 white cubes, it would be quite impossible to substitute the 13 dominoes.

Another little problem suggests itself in connection with those same 27 cubes. Say they are made of wood, and a termite starts tunneling from the center of an outer cube. Can he traverse all the cubes, going from center to center passing only through face-to-face connections, never in any *diagonal* direction, never entering the same cube more than once, and end up in the center cube? If he can do so, the determination of a successful route will prove it.

But, if you think he cannot do so, then prove the impossibility of such a path. The alternate coloration furnishes a clue to the solution to this puzzle.

Figure 8-5 shows a straight tromino, Figure 8-6 a monomino. It is obviously impossible to cover an 8 by 8 checkerboard with straight trominoes alone, for 64 is not a multiple of 3. But, can we cover an 8 by 8 checkerboard with 21 trominoes and 1 monomino?

To answer this, we color the checkerboard with three colors—shaded, white, black—as indicated in Figure 8-7, in which we have 22 shaded, 21 black and 21 white squares. Each of the 21 trominoes would have to cover 1 shaded, 1 black and 1 white square. A successful solution must depend on the placing of the single monomino, so we consider the various placings available for it.

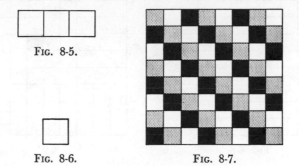

FIG. 8-5.

FIG. 8-6.   FIG. 8-7.

If the monomino is placed in the lower left hand corner, shown as black in the figure, there will remain 22 shaded, 21 white and 20 black squares: these obviously cannot be covered by 21 straight trominoes, because of the color disparity. We now invoke the useful principles of symmetry. Any one of the other three corner squares might equally well have been colored black, so by symmetry we can see that placing the monomino on any corner square will make it impossible to cover the rest of the board with straight trominoes.

If the monomino is placed on any white or black square, the resulting color disparity will preclude any possibility of covering the remaining squares with straight trominoes. And, again invoking symmetry, the same must apply if the monomino is placed on any square which is symmetric to a white or black square, i.e. a square which corresponds in its own half of the board to a white or black square in the other half.

Only four shaded squares of the 22 indicated in Figure 8-7 are not symmetric to white or black squares. These are shown in Figure 8-8.

Then, if the monomino is placed on any one of these four shaded squares, the remaining squares on the board can be covered with straight trominoes. This is the required solution, one arrangement being shown in Figure 8-9.

Many problems that involve shapes and diagrams can be handled more simply if colors are used. This has been shown in the two cases already discussed, and a further example will be found later in this chapter. We have touched on monominoes, dominoes and

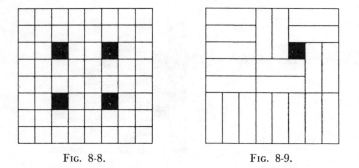

FIG. 8-8.   FIG. 8-9.

trominoes, however. And these lead to more complex shapes in the general category of polyominoes—shapes built up of squares connected edge-to-edge.

Figure 8-10 shows the first nine members of the polyomino family. The study of these and of the various configurations that can be built up by combining sets of them has developed into a most popular and diverting branch of recreational mathematics.

There are only two distinct types of trominoes, straight and right, and only five types of tetrominoes: with these alone only a few different configurations can be developed. However, there are 12 pentominoes, as shown in Figure 8-11.

There are 35 hexominoes, 108 heptominoes, and rapidly increasing numbers of distinct types in the successively higher orders of polyominoes—4466 dekominoes (10-square polyominoes). But pentominoes, with their 12 distinct types, are handy and have in fact

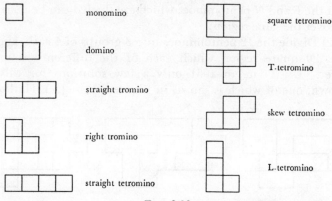

FIG. 8-10.

become more popular than any other polyominoes. Indeed, several manufacturers have marketed games based on the manipulation of pentominoes.

In Figure 8-11 a distinguishing letter is given for each different

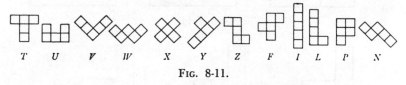

FIG. 8-11.

shape. These provide a ready means of identification and a mnemonic aid, and these letters can be remembered through a further mnemonic aid: *TUVWXYZ* and *FILiPiNo*, i.e. the last seven letters of the alphabet and the applicable letters of Filipino.

There are endless amusing problems relating to pentominoes. Here are some examples.

(1) Arrange the 12 pentominoes to form two 5 by 6 rectangles. The solution shown in Figure 8-12 is the only one known (except

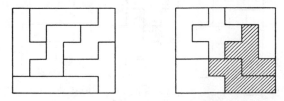

FIG. 8-12.

that the F and N pentominoes which are shaded can be rearranged
to cover the same region).

(2) Divide the 12 pentominoes into 3 groups of 4 each, and then
find a 20-square region which each of the different groups will
cover. Until late in 1961 only a few solutions for this were
known, one of which is shown in Figure 8-13. Jack Halliburton

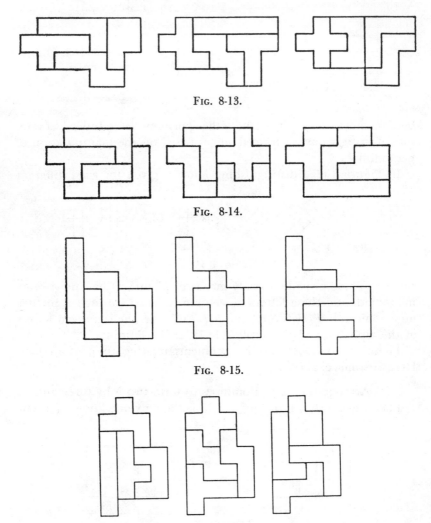

Fig. 8-13.

Fig. 8-14.

Fig. 8-15.

Fig. 8-16.

discovered another solution, shown in Figure 8-14. This particular problem posed in *Recreational Mathematics Magazine* stimulated other pentomino enthusiasts to such a degree that within a few months a total of 115 solutions was found! Two of these are shown in Figures 8-15 and 8-16.

In this connection we may mention the possible problem of trying to cover a 4 by 5 rectangle of 20 squares with each of 3 groups of the 12 pentominoes. Consider the 4 by 5 region shown in Figure 8-17. The X pentomino must be in one group and so must be used. In the figure, the X pentomino has been placed in the only position which does not leave isolated single squares (which could not be covered by any pentomino). Figure 8-17 shows that, with the

FIG. 8-17.

X pentomino in this position, we would have to duplicate either the U pentomino or the L pentomino in order to cover the remaining 15 squares. So we see that there is no solution for the 4 by 5 rectangle in this case.

Coming back to the original problem of the general 20-square region, for which some 115 solutions are known, let us consider what conditions must be satisfied for a solution to be possible. Any configuration of 20 squares, if it be a possible shape in which to place each of the 3 groups, must be such that each of the 12 pentominoes will isolate a number of squares that is a multiple of five when suitably placed.

For the second condition we use our coloration concept. If each of the 12 pentominoes is placed on a checkerboard, each will cover two squares of one color and three of the other, *except* that the X pentomino will cover four squares of one color and only one square of the other. So any four pieces, excluding the X, will cover at least 8 or at most 12 squares of one color. Without loss of generality we may assume that this one color will be black.

Now, a region that covers 8 black squares and 12 white squares of a checkerboard can be changed to a covering of 12 black squares and 8 white squares by merely switching colors. Similarly, a region covering 9 black squares can be transformed into a region with 11 black squares. Switching a region that contains 10 black squares does not result in any change in the number of black squares.

This means that a region, if there is a solution likely to be found, must be such that the *maximum* number of black squares covered by the region is 10, 11 or 12.   Figure 8-18 shows a region which can

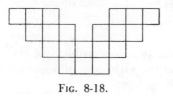

cover a maximum of 13 black squares. That no solution to the 20-square problem is possible with this region can easily be shown by noting that there is only one possible position for the *I* pentomino—and it isolates a region of 3 squares.

FIG. 8-18.

So we have the second condition that the possibly solvable configuration of 20 squares must include 10, 11 or 12 black squares if transferred to a checkerboard.

Unfortunately, although these two conditions are *necessary* conditions, some regions that satisfy them will have no solutions (e.g. the 4 by 5 region discussed above).   So far, no *sufficient* conditions have been discovered.

(3) Divide the 12 pentominoes into 3 groups of 4 each, and subdivide each group into 2 pairs of pentominoes.   Now find three 10-square regions, each of which can be covered (independently) by both pairs of pieces in the respective groups.   There are 18 known solutions, two of which are shown in Figures 8-19 and 8-20.   It has been proved that these constitute all the possible solutions.

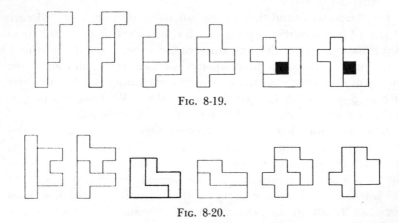

FIG. 8-19.

FIG. 8-20.

(4) Divide the 12 pentominoes into 4 groups of 3 each, and find a 15-square region which each of the 4 groups will cover. No solution to this problem is known, but neither is there any proof that it is impossible to solve.

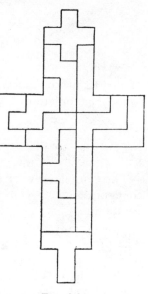

Another favorite type of pentomino pastime involves the problem of forming various shapes using all 12 pieces and, in some cases, auxiliary pieces outside the range of pentominoes. Figure 8-21 shows an elongated cross configuration which can be covered, in several different ways, by the 12 pentominoes.

Figure 8-22 shows a 61-square configuration, with a monomino in the center. It has been proven that it is impossible to cover the remaining 60 squares with the 12 pentominoes. For other positions of the monomino, however, there are some solutions: one of these is shown in Figure 8-23.

Figure 8-24 shows a 60-square con-

Fig. 8-21.

figuration that certainly might be expected to be solvable for the 12 pentominoes. It can be proved quite neatly, however, that no solution is possible.

Counting the squares that form the edges of this configuration we find a total of 22. The *maximum* number of squares that each

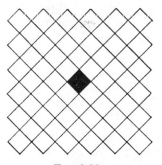

Fig. 8-22.

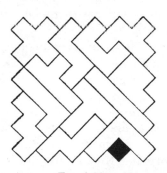

Fig. 8-23.

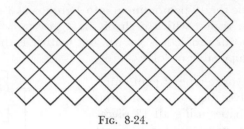

Fig. 8-24.

of the 12 pentominoes can contribute to an edge is given in the following tabulation:

$$T - 1 \qquad W - 3 \qquad Z - 1 \qquad L - 1$$
$$U - 1 \qquad X - 3 \qquad F - 3 \qquad P - 2$$
$$V - 1 \qquad Y - 2 \qquad I - 1 \qquad N - 2$$

These total only 21 possible edge-square contributions, so the configuration shown in Figure 8-24 cannot be formed with the 12 pentominoes.

Some proofs can be surprisingly simple, others very complex. The proof mentioned in connection with the configuration shown in Figure 8-22 involves two pages of preliminary reasoning just to establish the validity of the arguments to be used in the main proof that follows! And in this we can see why the study of polyominoes has its very rightful place in the broad field of mathematics.

The reader may feel tempted to delve more deeply into the possibilities with polyominoes, but we would like to pass on to some of the many problems that arise in connection with chessmen and chessboards.

In the chapter on *Mystic Arrays* (Chapter 3) we outlined general methods for constructing magic squares. There is a further method based on the moves of the chess Knight (a Knight moves from one corner of a 1 by 2 rectangle to the opposite corner).

In Chapter 3 the de la Loubère method for constructing odd-order magic squares was detailed. Surprisingly, the same principle can be used with Knight's moves: moving 2 squares up and 1 square across, instead of the one square up and one across detailed for de la Loubère's method. Again, if a move would take the Knight outside the magic square, we imagine the array extended and then move the Knight into the corresponding square in the original array. If a move would take the Knight into an "occupied" square, we drop one square—as in de la Loubère's method. The 5 by 5 magic square shown in Figure 8-25 illustrates the Knight's move method.

Regular chess problems, those end games in which white has to mate in a specified number of moves, provide endless diversion and frustration for chess addicts. But there are many interesting puzzles based on various moves that the chessmen are permitted to make.

10	18	1	14	22
11	24	7	20	3
17	5	13	21	9
23	6	19	2	15
4	12	25	8	16

Can you place the 16 white pieces on a chessboard so that each piece protects only one other piece? One solution is given in Figure 8-26, but there are other solutions.

Fig. 8-25.

How do you place the 8 senior chessmen (white or black, but no pawns) so that:

(a) the maximum number of squares are under attack
(b) the minimum number of squares are under attack?

Fig. 8-26.

FIG. 8-27.

FIG. 8-28.

It appears that no more than 63 squares can be attacked simultaneously in this way, but it has never proved that it is impossible to have all 64 squares under attack. B. R. Trone has shown that all 64 squares can be under attack, but with both bishops on the same color squares, making the solution invalid. In Figure 8-27 we do show one of several ways in which every one of the 64 squares is either under attack *or* occupied. And Figure 8-28 shows a 16-square solution to the minimum attack problem, with the two Bishops on different color squares, of course.

In a regular 8 by 8 chessboard, how many individual rectangles can you identify? How many in the general case of an N by N checkerboard?

Chapter 9

ALPHAMETICS AND THE LIKE

Nobody knows when or where letter-arithmetic started, although it seems certain that this form of puzzle was already well established in India and in China a thousand years ago.

Basically, such a puzzle presents the regular lay-out of a simple arithmetical calculation—addition, subtraction, multiplication, or division—but with the digits transposed into letters of the alphabet (or other non-numerical symbols). In the one puzzle each letter will stand for a particular but different digit. And the solver is required to discover the digital values of the letters, and so to reinstate the original calculation.

There were variations of course on the basic theme. For example, some of the digits might be indicated merely by asterisks without regard to their respective digital values.

In 1931 the designation "Cryptarithm" was first used for these puzzles, following on a suggestion by M. Vatriquant in the famous recreational mathematics magazine *Sphinx* (now defunct). But this entailed no change in the form of the puzzles which, except to the most ardent puzzlers, presented the somewhat uninspiring and forbidding appearance of an apparently meaningless jumble of letters.

This is exemplified in M. Vatriquant's puzzle which introduced the new designation "cryptarithm":

$$
\begin{array}{ccc}
A & B & C \\
  & D & E \\
\hline
F & E & C \\
D & E & C \\
\hline
H & G & B & C
\end{array}
$$

In 1955, J. A. H. Hunter coined a new word *alphametic* as a designation for a more attractive form of cryptarithm, in which meaningful words and even phrases would appear for the beguilement of the prospective solver. For example, the puzzle of M. Vatriquant could be more amusingly presented in true alphametic form as:

90

```
        B   U   T
            W   E
        ─────────
        G   E   T
    W   E   T
    ─────────────
    L   O   U   T    Note: See Answers section.
```

The new designation, for such meaningful-words puzzles, arose from a fortunate typographical error. A correspondent had written, asking J. A. H. Hunter to pose what he described as alphabetical puzzles more frequently in the latter's daily problem feature, at that time published only in the Toronto "Globe and Mail" and one other newspaper. The word *alphabetical* was repeated several times in the letter, and once it appeared as *alphametical*. And it was this that prompted Hunter to adopt and eventually popularize the designation *alphametic* and, with it, the modern meaningful-words form that has so greatly increased popular interest in these puzzles.

No rules can be laid down for the solving of alphametics or cryptarithms. Rarely is any mathematical know-how required beyond a complete understanding of the basic facts of simple arithmetic, coupled with much clear and logical reasoning, and often a fair amount of determination and patience.

The examples that follow are graded from easier up to more difficult, most of them being true alphametics. Space does not permit the inclusion of the full and detailed solutions, some of which are very lengthy, but for each of these puzzles the suggested method of solution is outlined in the appropriate section later in this book.

```
(1)     L   O   S   E          (2) D  O  )  F   L   Y   (  D  O
        S   E   A   L                       I   F
    ─────────────────                       ─────────
    S   A   L   E   S                       D   R   Y
                                            D   R   Y
                                            ─────────
                                            ─   ─   ─
```

```
(3)     S   E   T              (4) T   R   I   E   D
        N   E   T                  D   R   I   V   E
    ─────────────                  ─────────────────
    U   S   E                      R   I   V   E   T
                A
    ─────────────
    L   U   R   E
```

```
(5)    F  U  N         (6)             T  H  E
       I  N                      S  E  V  E  N
       *  *  *                   S  E  V  E  N
    *  *  *                   T  E  A  S  E  R
    F  A  C  T
```

(7)   "I'll have an invoice for these," Mrs Pothersniff declared, picking up her purchases. So Peter wrote one out for her.

Some customers might have been amused, but not so this pon-
derous female. Poor Peter!

Each capital letter in the invoice stood for a different figure,
and in the three words (describing the items) the corresponding
figures were used for their letters. What was the total amount?

```
L   40082    @      NC  ¢ .. .. N.NN
C   1637592  @       C  ¢ .. ..   .BL
S   1632     @   $P.NL    .. .. S.CK
                                 $E.EO
```

```
(8) I  S  )  T  H  A  T  (  I  T
             T  R  Y
             ?  ?  ?
             ?  ?  ?
             ─  ─  ─
```

```
(9) E  V  E  )  T  R  U  S  T  S  (  A  R  T
                *  *  *  *
                   *  *  *
                   *  *  *
                   ─  ─  ─
```

```
(10) P  O  S  H   (11) L  A  D  )  T  O  O  K  (  N  O
     C  H  O  P             K  I  T
     S  H  O  P             *  *  *
                              *  *  *
                              ─  ─  ─
```

(12)   A sad, sad story! But it happens every day. The same letter
values apply in both parts.

```
        I  C  Y          C  C  C
     R  O  A  D             N  O
        C  A  R          C  A  R
  S  K  I  D  S
```

(13) C A N) M A R L (I T
                C A N
                S A I L
                * * * *
                ─ ─ ─ ─

(14)   X M A S
            M A I L
         E A R L Y
      P L E A S E

(15)   No collection of alphametics would be complete without the
first known example of this special form of such puzzles. A true
alphametic in every sense, this well known puzzle appeared some
thirty years before that designation was introduced.

            S E N D
            M O R E
         M O N E Y   Note: See Answers Section.

    Now we give the sequel, as a rather more difficult alphametic,
there being no connection between the two so far as letter values
are concerned.

                     H E
            S E N T
                     H E
            S E N T
               T H E
               T E N
            T H E N

(16)   E E R Y
               O W L
            * * * * *
         * * * * *
         * * * *
         R R R R R R

(17) M A N) D I Z Z Y (M A N
                * * *
                * * D *
                * * R
                * * * Y
                * * * *
                ─ ─ ─ ─

(18)   Not only is this addition "doubly correct," but THREE is indeed divisible by 3 and SEVEN by 7.

```
    T  H  R  E  E
    T  H  R  E  E
          O  N  E
  ────────────────
    S  E  V  E  N
```

(19)   In this curious little puzzle all the digits, as indicated by the asterisks, are prime numbers. May we remind the solver that 1 is not a prime number according to the accepted definition.

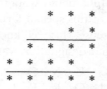

```
            *  *  *
               *  *
        ──────────────
         *  *  *  *
      *  *  *  *
      ──────────────
      *  *  *  *  *
```

(20)   In this variation on the alphametical theme, the letters of the five vital words (i.e. in capitals) conform to the normal rules.

A FOOT plus a FOOT plus a FOOT
Just totals a YARD, as you know.
But say, to a TRAY you add FOOD,
What total will give you a GLOW?

(21)
```
        F  O  U  R
           O  N  E
     T  H  R  E  E
     T  H  R  E  E
  ──────────────────
  E  L  E  V  E  N
```

(22)
```
           T  E  N
           T  E  N
        N  I  N  E
     E  I  G  H  T
     T  H  R  E  E
  ──────────────────
     F  O  R  T  Y
```

Note: FOUR is divisible by 4.   Note: NINE is a perfect square.

(23) * * *) * * * * * * * * (* 7 * * *
```
            *  *  *  *
          ──────────────
            *  *  *
            *  *  *
            *  *  *  *
            *  *  *
          ──────────────
            *  *  *  *
            *  *  *  *
          ──────────────
                5
```

(24)   There are no restrictions on liquor ads in Kalota. Here is one
       that appeared recently in one of that charmed island's news-
       papers, accompanied by a tempting picture of a long cool drink.
       Indeed an alphametic in quite unusual form, with each capital
       letter standing for a different figure of course.

> "SPLASH added to this great
> SCOTCH will produce a POETIC
> result, giving a total of but
> two perfect cubes if you take
> the ICE from the TOP."

(25)   Finally, we conclude with this rarity.   No digits at all are
       given, and there are no special conditions.   Not more than four
       or five puzzles have ever been produced to meet that specifica-
       tion.
       Note the decimal point in the quotient.

```
* * * ) * * * * * * ( * * * * * · * * * *
        * * *
        —————
        * * *
        * * *
        —————
          * * *
          * * *
          —————
            * * *
            * * *
            —————
              * * * *
              * * * *
              ———————
              — — — —
```

Chapter 10

WHAT ARE THE ODDS?

Before discussing probabilities and odds we must define each of these terms very clearly, for they are in no way synonymous.

The probability of a particular outcome from a trial is the ratio of the possible number of such particular outcomes to the total of all possible outcomes from the trial. For example, if we toss a coin once there are two possible outcomes: it may land heads up, or tails up. The probability that it will land heads up, therefore, is $\frac{1}{2}$. Similarly, if we throw a single die with 6 faces numbered 1 to 6, the probability of 5 appearing will be $\frac{1}{6}$: the appearance of the 5 is one only amongst the 6 possible outcomes from that throw.

The odds *in favor of* a particular outcome resulting from a trial is the ratio of the possible number of such particular outcomes to the total of all possible *other* outcomes from the trial. In the simple case of our coin, the odds in favor of heads (or tails) will be 1:1, which we describe as even odds. With our single die, the odds in favor of throwing a 5 in one throw will be $\frac{1}{5}$: the appearance of the 5 is one outcome, as against the five other possible outcomes.

We often refer to the odds *against* some particular outcome. The odds *against* a particular outcome can be found by taking the reciprocal of the odds *in favor*. For example, the odds against throwing a 5, with one throw of our single die will be 5/1, or five-to-one against.

If we throw the die six times, the 5 might be expected to appear once. There could be no certainty of this, however. If the die were thrown six thousand times, it is true that the 5 would appear approximately one thousand times: with six million throws it would appear approximately one million times, the approximation being closer proportionately than for six thousand throws. Probability provides a measure of the likelihood of an event happening, or not happening, but cannot be used for making an exact forecast of what will happen in any finite number of trials.

We can calculate the probability of the 5 appearing *at least once* in six throws of the die. The simplest approach is to calculate the probability that it will not appear at all in six throws.

The probability that the 5 will *not* appear at the first throw is obviously $\frac{5}{6}$. The probability that it will not appear at the second throw is also $\frac{5}{6}$. So the probability that it will not appear in the first two throws is ($\frac{5}{6} \times \frac{5}{6}$), i.e. 25/36. The probability that it will not appear in the first six throws is ($\frac{5}{6}$)^6, i.e. approximately $\frac{1}{3}$. Hence, the probability that it *will* appear *at least once* in six throws is approximately $\frac{2}{3}$. Very roughly, then, we can say that there is a better than even chance of the 5 appearing *at least once* in six throws.

What is the probability of having heads appear with each of two successive tossings of a coin? The probability of heads on each separate occasion is $\frac{1}{2}$, so the probability that it will appear on *both* occasions is ($\frac{1}{2}$)^2, i.e. $\frac{1}{4}$. The same probability applies, of course, for tossing tails on *both* occasions. These calculated probabilities can be checked very easily by listing the possible outcomes from tossing a coin twice in succession:

Heads, tails.   Heads, heads.   Tails, tails.   Tails, heads.

There are four different outcomes, each equally likely. Only one gives successive heads (i.e. probability $\frac{1}{4}$), and only one gives successive tails (i.e. probability $\frac{1}{4}$).

If you tossed a coin six times in succession, with heads each time, you might well think that the probability of having heads with a seventh toss was very small. But, assuming there was nothing wrong in the coin itself, the probability of the seventh toss giving heads or tails) would still be $\frac{1}{2}$, i.e. an even chance.

On the other hand, the probability of tossing heads seven times in succession, is indeed very small. It is, in fact, ($\frac{1}{2}$)^7, i.e. 1/128.

This last example shows how important it is to define the problem precisely when considering any calculation of probability. One must also be quite sure as to the validity of any assumptions that may be made, as for example an assumption that various outcomes are equally likely.

The evaluation of π has been carried to more than 100,000 decimal places. It might seem reasonable to assume that each of the ten digits, 0, 1, 2, 3, 4, 5, 6, 7, 8, 9 is equally likely to appear in that sequence of more than 100,000 digits. Indeed, a count of the digits does confirm this. However, it has never been proved that this assumption is truly valid. There must remain the possibility that, if the evaluation of π were carried to 200,000 decimal places, a preponderance of one or more digits might appear. Hence, for example, we can only conjecture that the probability of 6 appearing considerably *more* than 10% in an extended evaluation of π is infinitesimal.

Many paradoxes have been developed in connection with even the simplest of probabilities. Some depend upon fallacies that are obvious, some on fallacies that can be demonstrated only by very advanced mathematical theory, and some have never yet been fully rationalized.

Lewis Carroll proposed an amusing paradox which has puzzled many an expert, although the fallacy can be pin-pointed without much difficulty. "A bag contains two counters, as to which nothing is known except that each is either black or white. Ascertain their colors without taking them out of the bag."

Carroll *proved* that the answer *must* be "1 white, 1 black," using the following reasoning:

We know that, if a bag contained 3 counters, 2 being black and 1 white, the probability of drawing a black counter would be $\frac{2}{3}$; and that any *other* state of affairs would *not* give this probability.

Now the probabilities, that the bag contains (a) BB, (b) BW (c) WW, are respectively $\frac{1}{4}, \frac{1}{2}, \frac{1}{4}$, assuming only the two counters.

Add a black counter to the two counters in the bag.

Now, the probabilities that the bag contains (a) BBB, (b) BWB, (c) WWB must be, as before, respectively $\frac{1}{4}, \frac{1}{2}, \frac{1}{4}$.

Hence, the probability of now drawing one black counter *from* the bag, must be:

$$(\tfrac{1}{4} \times 1) + (\tfrac{1}{2} \times \tfrac{2}{3}) + (\tfrac{1}{4} \times \tfrac{1}{3}) = \tfrac{2}{3}$$

But we have seen that this probability requires the bag to contain 2 black counters and 1 white counter, so the bag must contain BBW. Hence, before the one black counter was added, the bag must have contained one black and one white counter. A quite amazing, but apparently logical conclusion!

In fact, Carroll introduced his third counter merely to confuse the issue, for the same quite fallacious result can be derived without this complication.

With only the two counters, the probabilities that the bag contains (a) BB, (b) BW, (c) WW, are respectively $\frac{1}{4}$, $\frac{1}{2}$, $\frac{1}{4}$ as correctly stated in Carroll's argument. The probabilities of drawing one black counter in these respective individual cases are 1, $\frac{1}{2}$, 0. Combining these probabilities, the probability of drawing one black counter *from* the bag is:

$$(\tfrac{1}{4} \times 1) + (\tfrac{1}{2} \times \tfrac{1}{2}) + (\tfrac{1}{4} \times 0) = \tfrac{1}{2}$$

But, if the probability of drawing a black counter is $\frac{1}{2}$, the bag must contain one black and one white counter. This is the same as Carroll's conclusion, but reached without the use of a third counter. And we can now indicate the fallacy!

The fallacy lies in the final step in the argument. The probability of drawing one black counter from the bag is certainly $\frac{1}{2}$, as would be the probability of drawing one white counter from the bag. But this proves nothing at all as to the color of the remaining counter! We only know that each of the two counters is either black or white, so we must be just as likely to draw a white counter from the bag as a black counter. No amount of complicated calculation and reasoning can alter that fact. Hence no valid conclusion could be drawn from the confirmation of a probability that was already obvious and inherent in the problem as given.

Some of the more difficult paradoxes involve the concept of infinity. For example, the largest known prime number is $2^{44497} - 1$ (a number of 13395 digits!). This is a fantastically large number, but nevertheless the number of known primes is finite. The total number of primes, however, is infinite. So the probability that any particular prime number is a known prime must be zero. That implies quite definitely that it is impossible for any prime number to be known, from which we must conclude that no prime numbers are known!

The fallacy in this prime number paradox lies in the erroneous assumption that "infinitely small" is synonymous with "zero." We should have argued that the probability of any particular prime number being known is infinitely small.

One interesting geometrical probability paradox involves an equilateral triangle inscribed in a circle. What is the probability that a chord, drawn at random in the circle, will be longer than the side of the inscribed equilateral triangle? There are three classic approaches to this problem, each leading to a different evaluation of that probability, and all based on arguments that seem equally valid: and we here propose yet a fourth approach that leads to a fourth apparently reasonable value! These four different solutions will be outlined in turn, the classic three being taken first.

Figure 10-1 shows an equilateral triangle inscribed in a circle, with chords drawn perpendicular to the diameter that meets a vertex of the triangle. It is not difficult to prove that the distance from the center of the circle to any side of the triangle equals half the radius of the circle. Hence, in Figure 10-1, OC is half the radius OA, so OC is one fourth the diameter AB. Then, if we take the point D, such that $OD = OC$, it is obvious that any per-

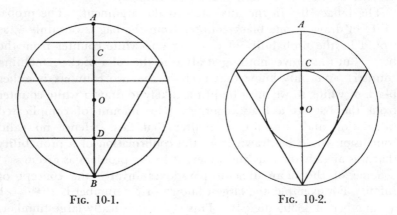

FIG. 10-1.   FIG. 10-2.

pendicular chord between C and D must be longer than the side of the triangle. Of all the perpendicular chords that can be drawn between A and B, just half would be between C and D. Therefore the required probability is $\frac{1}{2}$. The same argument would apply for any other orientation of the inscribed triangle within the circle, so we appear to have proved that the probability is indeed $\frac{1}{2}$.

Figure 10-2 illustrates the second approach. Here a circle has been inscribed within the equilateral triangle. As before, OC is half the radius OA of the original circle, hence the radius of the inner circle is half the radius of the outer circle. We now consider the mid-points of all possible chords that can be drawn for the original circle. Chords with mid-points within the inner circle must be longer than the triangle side: those with mid-points not within the inner circle must be shorter than the triangle side. So it would seem reasonable to take the probability that a random chord is longer than one side of the triangle as being the ratio of the area of the inner circle to the area of the outer circle. From the known ratio of the respective radii, the ratio of areas is 1 to 4, hence the required probability is $\frac{1}{4}$.

The third approach, a very simple one, is illustrated in Figure 10-3. Typical chords have been drawn from one vertex of the equilateral triangle. Obviously an infinite number of such chords can be drawn from that vertex, only those that pass through the triangle being longer than the triangle side. Since the chords can be drawn through a 180° range, within which a 60° range will

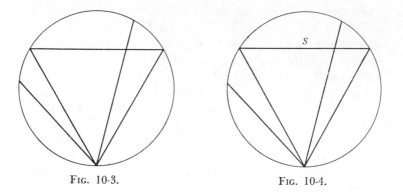

FIG. 10-3.   FIG. 10-4.

include all that pass through the triangle, it would seem that just one-third of them must be longer than the triangle side. This makes our required probability precisely $\frac{1}{3}$.

Finally we outline our own possible fourth approach to the problem, as illustrated in Figure 10-4. Here again we consider possible chords that can be drawn from one vertex, only those that pass through the triangle *and* the segment S being longer than the triangle side. An infinite number of chords can be drawn, and it seems reasonable to assume that an infinite number of chords must *completely* cover the entire area of the whole circle. Since none of an infinitude of different chords will overlap we take the ratio of areas as being identical with the ratio of the number of chords that traverse these areas. A simple calculation will show that the ratio of the area of the triangle *plus* the area of segment S, to the area of the whole circle, is approximately 14 to 23, say roughly 3 to 5. Hence we might conclude that our required probability is approximately $\frac{3}{5}$.

So, using four different approaches, we have derived four very different values for that probability: $\frac{1}{2}$, $\frac{1}{4}$, $\frac{1}{3}$, and finally the approximate $\frac{3}{5}$. Each approach involved an assumption that may or may not have been valid, although each appeared reasonable. The actual probability in this problem must be some definite and fixed ratio, so at least three of those approaches must have been based on false assumptions: possibly all four. But who can decide?

In most practical cases, however, probabilities can be calculated on assumptions that are obviously valid, or that can be proved valid. And some of the results can be surprising!

Say there are about 25 to 30 people in a room. What is the probability that at least two of them have the same birthday in the year (i.e. same date in the year, not necessarily the same year)? That probability could be expected to be very small, and indeed many people would gladly take you up on a small wager if you bet that at least 2 had the same birthday. In fact, in any random assembly of more than 22 persons, it is better than an even chance that at least two of them will have the same birthday.

The exact probability, for an assembly of any particular number of people, can be calculated precisely. If we ignore February 29th, the possible Leap Year birthday, the precise probability is given by the formula:

$$1 - \frac{365!}{365^n \times (365 - n)!},$$ where there are n people.

The calculation of approximate values is quite simple, using logarithms. Typical values are listed:

People	Probability that 2 at least will have the same birthday
10	0.117
20	0.411
23	0.507
30	0.706
57	0.990
366	1.000

In that formula we ignored February 29th as a possible birthday. On that assumption, therefore, it would obviously be a certainty (i.e. probability = 1) that an assembly of 366 or more persons must include at least two with the same birthday. Taking February 29th into account, absolute certainty would require 367 or more in the assembly. But note that the odds are slightly better than even with an assembly of only 23 persons!

This birthday phenomenon can be tested without the probable embarrassment of checking birthdays at the next party you may attend: some ladies might suspect your motives! You might pick names in a biographical dictionary, and check for duplication of birthdays. For example, the list of the first 34 U.S. presidents shows that James K. Polk and Warren G. Harding were both born on November 2nd.

It must always be remembered that a correctly calculated probability will not necessarily provide a basis for a sure and accurate forecast of future events. Only a probability of unity, or conversely of precisely zero, can imply absolute certainty. In the more important uses of probability, however, the theoretical values are applied in considering large numbers of individual events, and with these the known probabilities can provide forecasts that will approximate quite closely *on the average* to actual future happenings.

Life insurance premiums are based on the mortality records of millions of people. From such records an actuary can calculate the probability that one individual will, for example, survive to receive personal payment on an endowment policy 30 years hence. Premiums will be based on the near certainty that the *average* experience for the future, with a very large group of similar cases, will be in close agreement with what the probability considerations forecast. The *individuals* in that group, however, can be expected to differ widely and in varying degrees from the average experience. The insurance company will lose heavily on some of those individuals, and will profit handsomely on others. But, provided the premiums have been calculated correctly, the overall experience with that group will leave a small and reasonable margin of profit for the company.

Probability has become very important, too, in the field of atomic physics. The movements of incredibly minute particles, such as protons, cannot be forecast individually. But here the physicist is dealing with enormous assemblies of individual particles, and the *average* behavior of those particles can be forecast with reasonable accuracy by proper application of the theory of probability.

All of that is somewhat remote from recreational mathematics! But the whole theory of probability has been developed from the question that Pascal asked about 300 years ago: "What are the odds?"

Chapter 11

STORY TEASERS

1. NOT A CARE IN THE WORLD

"You're poor, Mike O'Hara,"
                                        his visitor said.
"An acre of land and
                        a cow and a shed."
"I'm rich," replied Michael,
                        content with his lot.
"In Ireland there isn't
                        the like of my plot.
"Its sides are all equal:
                        just three sides, not four.
"The beast grazes half and
                        she doesn't need more.
"A stake in one corner
                        with just enough rope,
"And grass and some clover:
                        she's plenty of scope.
"We're happy." He smiled. "I've
                        enough and I'm free."
So tell us how long the
                        cow's tether would be.

Note: No detailed solution will be given for this.

105

2. A GAME OF MARBLES

"Help yourselves, but only twelve each," said Jim grabbing a dozen marbles from the box. "We've less green than blue and less blue than red, so each take most red and fewest green."

Jim having done so himself, the other boys complied. There were only the three colors, and the box contained just the right number.

"We've all picked different selections," remarked Joe. "I'm the only one with four blues."

"So what?" Pete picked up one of his green marbles that he'd dropped. "Let's play."

And play they did.

There were twenty-six red marbles altogether. So how many boys were there?

3. THE BONUS

Jack smiled as his secretary came into his room. "I'm through now, Betty," he said. "Please call in the others."

Very soon the whole staff, only five including Betty, were there before him wondering what it was all about. But their boss quickly put them at their ease. "You'll be happy to know that I've pulled off the Cramond deal at last," he told them. "And here's a bonus of $260 amongst you to mark the occasion."

Betty likes to think herself more junior executive than secretary. "Maybe he won't count me in with the clerks." The chilling thought hit her.

But all was well, for Jack went on. "I've figured it out exactly in proportion to completed years you've each been with me, but allowing half as much again for a man as a girl." With that he handed each an envelope, somewhat embarrassed by the outburst of thanks this evoked.

It was a good break for all of them, their completed years of service being two, three, five, six, and seven respectively. But you'll have to figure out the amount that went to the female side of Jack's little staff.

4. THE JOY-RIDE

"You're in a rush, Bill," remarked the professor, as his friend gulped down the last of his coffee and stood up to go.

"Taking three girls for a drive," replied the other.

The professor laughed. "So that's it! How old might the three graces be?"

His colleague thought a moment. "Multiplying their ages together you get twenty-four hundred and fifty, but their sum is just twice your own age."

The professor shook his head. "Very clever, but that still leaves their ages in doubt."

Bill was with him there, however. "Yes, I forgot to mention that I'm at least a year younger than the oldest," he said. And that made everything clear.

The professor knew his friend's age, of course. What do you make it?

5. WOULD THEY MEET?

"Where are you calling from?" asked Bert, watching the jam of traffic outside his office on the corner of Merton and Spruce.

"A pay phone on Dale and King," came Ben's faint reply. "Four blocks south and a few blocks east from you."

Bert glanced at the clock. "You start walking," he shouted, "and we'll meet halfway." He banged down the receiver, and only then did he realize what he'd done.

There are, in fact, just seventy different regular routes between the two intersections and with nothing to choose between them as regards distance.

So how do you interpret Ben's word "few"?

6. AFTER THE PARTY

"They seemed sober enough when they left last night," said Bob, just back from the office.

"No worse than you," his wife assured him. "But why?"

Bob smiled wanly. "They've been calling me all day, all four of them," he told her. "And I had to straighten out the tangle. It seems each took the wrong coat and hat, one man's hat and another man's coat."

"I'm surprised that you got home at all." Betty laughed. "Go on with the sad story."

"Well, I sorted it out. Joe took the coat belonging to the guy whose hat was taken by Steve, and Steve's coat was taken by the character who took Joe's hat."

"What about Ron?" Betty was enjoying this.

"He called first," replied Bob. "He'd taken Doug's hat."

It must have been quite a party! Whose coat and whose hat had Joe and Steve taken?

7. A MILD GAMBLE

"I've no pennies," said Hank, jingling his cash. "Have you any?"

Ben checked. "Just five," he replied. "But why?"

"Just wondered. Then we'll make it a little gamble." Hank started dealing the cards. "First game the loser loses a fifth of his money, the second game a quarter of what he has then, and the third game the loser pays a third of whatever he may have."

So they played and paid. But, after paying his loss on the third game Ben stood up. "I'm through," he declared. "Too much work for too little, and now there's only seven cents difference between us."

It was certainly little, since they had only seventy-five cents altogether. But how much did Hank have when they started?

8. A FRIEND IN NEED

Holding a match to his cigar, John leaned back in his chair very satisfied with life. "Yes," he chuckled. "I never thought to have it so good when we were teenagers together thirty years ago."

His visitor smiled. They had been good friends in those days, but that was long ago. What was an old friendship worth today in terms of the job he needed so badly? "What about your two brothers?" he asked. "Younger than you, weren't they?"

John nodded. "Doing well. Ben, that's the youngest, seems near to his first million. And Ted's one of the few bright boys at Washington. But you were great on figures, Bill. So what about this?"

For a few moments the tycoon scribbled, while Bill waited hopefully. "Ben's age, multiplied by the difference between Ted and myself, comes to one year less than my age multiplied by the difference between their ages. And I've taken our ages in full years."

"Too bad." Bill shook his head sadly. "I planned to ask you for a job, but I guess you've learned to handle your own figuring!"

He got the job, so it may amuse you to discover what those three ages were.

9. HIS FIRST JOB

"Hi, Johnnie," cried Joe, meeting the youngster on the street one Sunday. "Haven't seen you for quite a while, but I heard you'd started work."

"Some weeks ago," replied Johnnie. "It's piece work and I'm doing fine. My first week there I made more than forty bucks, and each week since I've earned just ninety-nine cents more than the week before."

Joe smiled. "That's a coincidence," he commented. "Let's hope you'll continue that way."

"Well, I guess I'll soon make better than sixty a week," the boy told him. "So far I've earned exactly $407 since I started, and that's not bad."

How much had he earned his first week?

10. TO THE COTTAGE

"I took the family to the cottage," said Bob. "It's nice. So quiet with no cars honking at night."

"Your cop's on the job, then," Len commented. "If you have one, that is."

"We don't need cops." Bob laughed. "But there's a teaser for you in our drive up. The first fifteen miles we only averaged forty. The next few ninths of the total distance was quite a bit faster. That left a seventh of the total distance, and I did it faster still. It's well over a hundred miles to the cottage, and I averaged just fifty-six the whole way."

"What d'you mean by 'a few ninths'?" Len asked.

"Exact whole number of ninths," his friend replied. "And for each of those last two stretches I averaged whole numbers of miles per hour."

Bob would not risk really crazy speeds with his family in the car, but maybe it's as well there were no cops up there! What was his average speed for that final seventh of the trip?

11. LEFT! RIGHT! LEFT!

"Dad!   Look at the soldiers,"
            cries Pete as they pause.
"There's one man left out
            when they line up in fours."

"He'd still be left out if
            they formed up in threes,"
His father declares and
            of course Pete agrees.

"He's right out of luck, for
            without any doubt
"If they formed up in fives he
            would still be left out."

There aren't many soldiers
            at drill in the Square:
Not even a hundred.
            How many were there?

Note: No detailed solution will be given for this.

12. THAT ENGLISH WEATHER

"So you had about five weeks over in England," said Bill, offering his friend a cigarette. "How was the weather?"

"Just as they say, more wet days than dry." Doug laughed. "And that reminds me. You know I'd nearly quit smoking. Over there I found myself smoking an average of twenty cigarettes a day the days it rained, but each dry day I smoked only a fifth of my average for the whole time there."

That's what a climate can do! But how many days did it rain during Doug's visit to the Old Country?

13. ONE CENT UP, ONE CENT DOWN

"You were quick," exclaimed Susan, when her husband came into their little shop. "Didn't you see anything?"

"Plenty, and for plenty of money," replied Ben. "But I did buy about six dozen blouses."

"We can sell them if the prices are right," Susan commented. "How much were they?"

Ben searched his pockets. "I must have left the invoice in the car," he said at last. "I bought three styles, each lot with one more blouse in it than the next. They were priced low, so no difference whether one bought singly or in quantity, but there was a two cents difference between the dearest style and the cheapest. Thirty-two dollars forty-five altogether."

Susan was puzzled. "I don't get it," she told him. But just then a customer came in.

Can you figure out the details?

14. FOUR GENERATIONS

He had been in the living room many times, but always with
Ted. Waiting now for his friend to come down, Ken had a chance
to look around. "Who are they?" he asked, pointing to a framed
photo of two men.

"My wife's father and grandfather," John told him. "Taken this
year."

"Ted's great grandad! Whatever age must he be?" exclaimed
the boy. "I never guessed Mrs Ryle was that young."

John was amused. "Well, I'll answer your question," he said.
"His age is what you get if you multiply her Dad's age by the
reverse of my wife's age, and then take away her age multiplied
by the reverse of her Dad's age."

"You mean the figures of the age in reverse order?" Ken had
been listening carefully.

"That's right." John nodded. "And my wife's age and her
father's age add up to six years less than her Grandad's."

Ted came into the room just then and rushed his friend out of
the house.

How old was his mother?

15. IN THE HOBBY SHOP

"I was in Sam's store today," said Jim. "He told me the kids had
been in earlier for a model airplane kit. Thirteen dollars fifty it
cost them."

"Well!" his wife exclaimed. "Where did they ever get the
money?"

"It was all in coins from their piggy banks, and Sam was still
laughing about it," Jim replied. "Ken had coins all of one de-
nomination, and Pete three times as many but all of another. It
seems the kids arranged all the coins in equal piles, as many rows
of piles as there were piles to a row, and as many piles to a row as
there were coins to a pile. That's how Sam put it."

"I guess they had more fun watching him check the cash," Pam
commented. "There must have been a lot of coins."

How many coins would you say?

16. ONE UP ON THE JONESES

George was looking pleased with life when he came home. "Maybe you'll soon have a swimming pool," he said. "The first in this section."

"That's wonderful!" Mabel was thrilled. "How big?"

"Nothing is fixed yet, but I like Sam's quotation even if it is unusual," her husband replied. "He'll do it complete at $35 for every foot of width, plus $27 for every foot of length. Any fraction of a foot counting as a foot."

"Sounds odd," commented Mabel. "It will be the regular shape with square corners?"

"Sure," George told her. "We'll have the biggest area possible for less than $1000."

Let's hope Mabel wasn't disappointed. What would the dimensions of the pool be?

17. THE CLOCK

It's restful sitting in Tom's cosy den, talking quietly and maybe sipping a glass of his madeira.

I was there one Sunday and we had the usual business of his clock. When the radio gave the time at the hour, the ormolu antique was exactly three minutes slow.

"It loses seven minutes every hour," my old friend told me, as he had done so often before. "No more and no less, but I've gotten used to it that way."

When I spent a second evening with him later that same month, I remarked on the fact that the clock was dead right by radio time at the hour. It was rather late in the evening, but Tom assured me his treasure had not been adjusted or fixed since my previous visit.

What day of the week was this second visit?

18. ONLY SMALL CHANGE

"I'll pay," said Peter, glancing at the check.

"Not your last day in Kalota," said his companion, putting a handful of coins on the table. "You paid when I was in the States last month."

Peter smiled. "Okay, I won't argue. But I see you still have some American money."

"A few dimes, not two dozen altogether," Kiko told him. "Perhaps you will change them for me."

"Be glad if I can," Peter replied. "The rate is nineteen kuks to our dollar today, so let's see what we've got."

They emptied their pockets, and then Kiko laughed. "We both seem poor," he said. "You have two-thirds as many coins as me, and altogether we have only eight dollars value and all in dimes and kuks."

He was right, and neither had any paper money on him. Peter, in fact, did not have quite enough to give proper change for Kiko's dimes.

What coins did each of them have?

19. THE JOINT BIRTHDAY

"Got to go early," said Bob, bundling papers into a drawer.   "It's my wife's birthday and also my boy's."

Dick looked up from his crossword.   "That's odd," he commented.   "Both the same date.   How old's the kid now?"

"Try calling him that!"   Bob laughed.   "His mother's age and his, multiplied together, come to one more than the square of their difference."

Dick found this tougher than his crossword puzzle.   What do you make those two ages?

20. AT THE ART STORE

Sam was busy when Joe entered his little store, so he stood waiting a while.   And very soon the customer seemed satisfied.   "That'll be all," he declared, handing over some money.   "Just those five, and I'll expect them tomorrow."

"That was quite a sale," chuckled Sam, turning to Joe as soon as the man had gone.   "One cent over a hundred dollars."

"A hundred dollars?"   It didn't make sense.   On the counter there lay five picture-frames of various sizes, their respective prices marked clearly on the glass: $2.10, $3.30, $4.62, $7.70, and $11.55. "It looks more like thirty to me," Joe commented.

Sam laughed.   "Those are what he bought," he said, nodding towards the frames, "but more than one of each, and no discounts."

Another customer appeared at that moment, so Joe left Sam to his prey.

What were the details of that purchase?

21. MAGNA CARTA

"Bring sack!" they cry,
    "Bring wine, bring mead!
The Charter's signed,
       and all's agreed."
King John with all
       his barons drinks;
Each once with each
       his goblet clinks.
A bell, each goblet
       peals a chime
To herald freedom
       in their time.
That's what they say.
       'Twas long ago,
But here there's something
       we should know.
Those clinks all totalled
       nine nought three:
How many goblets
       there d'you see?

Note: No detailed solution will be given for this little fantasy.

22. DO IT YOURSELF

Bill was too much absorbed in his work to greet his visitor right away. "Trying to smarten up this old table," he said, carefully placing a square tile on the cemented surface. "It's an idea Helen got from some magazine."

Tom watched critically for a few moments and then jotted down some figures on a scrap of paper. "Wouldn't smaller tiles look better?" he suggested. "If they were just three-quarters of an inch less each way, I figure you would need two hundred and fifty more to cover the top but it would be well worth it."

"Two more than that." Bill looked up, shaking his head. "But it's not possible with this sort of tile, as they only come in exact inches."

So that was that! And anyway it would have meant more work.

What size tiles was Bill using?

23. WHO PLAYS SWITCH?

"Light drizzle turning to snow," the weather man had said, but the drizzle wasn't even light.

"We can't go out. That's for sure," grumbled Jim, looking around at the other kids who had come in for his birthday party. "What say we play switch?"

So cards were found and they all sat at the big table. "Last time you stopped when you'd won three games in a row," Joan reminded her brother. "Let's agree to stop only as soon as everybody has won at least once."

"Okay by me," declared Jim. "Each game the losers all put a nickel in the kitty, and the winner takes out twenty-five cents. But the winner of the last game clears the kitty."

This seemed a good idea, so they started. And Jim's birthday luck didn't desert him. He won only one game, but it was the last and he ended up exactly a dollar to the good.

How many games did they play?

24. A TALE OF SOME CAKES

"I put the cakes in the kitchen," said Garry. "Three different sorts. 12¢, 14¢, and 17¢. Just two dollars altogether."

"That's fine," his mother declared. "How many did you buy?"

The boy told her the total number, and his mother went on reading. But some moments later she stopped, and then did some figuring on a scrap of paper. "I still can't tell how many you got of each," she said. "Did you buy only one of one sort?"

Garry answered her question. Only one word, but enough to clear up any doubts as to the details of his purchase.

How many had he bought at each price?

25. THE ROAD WENT THROUGH

"Well, we're all set to start," Steve declared. "Six hundred men altogether, including foremen. They hail from three different provinces, so let's hope there won't be much fighting."

"A tough lot." Bill nodded. "They'll be divided up, though."

"Bet your life they will!" Steve laughed. "As far as possible, the men of each province are split up into sections of sixteen plus a foreman from that province to each section. Then all the remaining men are in two equal smaller sections, one all from Manitoba including the foreman, the other all Ontario men with a Quebec foreman."

"Sounds like our Army days," commented his friend. "Except that there's a meaning to your scheme."

"You've got something there," Steve chuckled. "But luckily there are the same number of foremen from each of those three provinces."

How many men, and how many foremen came from each province?

26. A BREAK FROM READING

Paul put down his book and did some figuring on a scrap of paper. "There's a lot of reading in this," he told his wife. "I think you'll enjoy it too."

"Don't keep me waiting too long then." Joan smiled. "How much more have you to read?"

"Nearly a hundred and fifty pages," replied Paul. "But I've just seen something quite odd about that. The first chapter starts at page thirteen, and the numbers of all the pages I've read before the page I'm on now add up to the same as all the pages I'll have to read after this one."

"What about the pages you skipped?" asked Joan. "Anyway you can figure those things out when you're through. I want the book."

In fact he had not skipped any pages. What was the number of the page he was reading?

27. HIS PRIVATE ARMY

Peter's quite crazy about soldiers, and today his complete army was set up on the big table in two perfectly arranged solid squares.

"An amazing collection," his father commented, eyeing the formations quizzically. "But I see no officers."

"I'm in charge, Dad," the boy explained. "One square's got three more to the side than the other, and now I'm going to switch them all into five squares all exactly the same."

"That'll keep you busy." His father laughed. "You must have at least a thousand there."

Peter grinned, saying nothing. He knew he had far more than that.

How many did he have?

28. THE GREAT BALL OF KAAL

And Knok entered into the temple and spake unto Kylis, saying. Behold now that great ball of gold that thou hast made solid and true from the offerings of the people. Twenty kebals in diameter is it, and verily has its fame spread throughout the land.

Take then that ball and from all of it make three balls of gold solid and true, and no gold shalt thou waste in the making. And each diameter shall be a whole number of kebals, and thirty-eight kebals shall be the measure of the three taken together.

And Kylis did as he was bidden, and it came to pass that the greatest of the three balls of gold was taken to the temple of Kaal, where it rests to this day.

That's what the old book says, but what were the diameters?

29. AFTER THE RUSH

The little store was quiet when Dick went in, very different from just before Christmas. "You sold all your birds then," he remarked. "You did tell me you hadn't got in many this first time."

"I had only twenty altogether." The old man smiled. "They cost me just a hundred bucks, which wasn't too bad when I had to pay seven dollars each for the turkeys."

Jake does well even in that rather poor section, but this had been a new venture for him. "I remember you had quite an odd selection," Dick told him. "Only turkeys and geese and ducks, but I forget the numbers."

"Twice as many turkeys as geese, and as many ducks as I had to pay dollars for one goose and one duck," said Jake. "Same prices whether I got one or more of either or both."

How many of each had the old man bought?

30. WHOSE HAT?

It isn't all fun being Secretary of a Club, and that is what Mike was thinking after a morning on the telephone clearing up that stupid business of the hats.

It was bad enough to have four members complaining that they had lost their homburgs: not one of them had the courtesy to apologise for taking someone else's, even though it was by mistake.   But what really got the harassed Secretary was that Andy and Bill were accusing each other of downright theft, although neither had taken the other's headgear.

"Anyway," Mike consoled himself, "it's lucky only those four members were involved."   And it was also lucky that the other two, Don and Charlie, treated the whole thing as a joke: in fact, they both congratulated Mike on tracing those almost identical homburgs.

Charlie had taken the hat of the man who had not taken Charlie's hat, but who had taken the hat of the man who had taken the hat of the man who had been the first of the four to leave the Club the previous evening.   And Don had taken the hat of the man who had taken the hat of the man who had taken Andy's hat.

It was all very complicated, but perhaps you can say whose hat Charlie took.

31. ALL PASSENGERS ASHORE

"I'm sorry, dear lady,
                your landing's delayed.
"They only just told me
                your bar bill's unpaid.
"You paid up a dime less
                than half," said the purser.
"The cents and the dollars
                you switched vice versa."
How much was still due from
                the boozy old maid?

Note: No detailed solution will be given for this.

32. THE DOLLAR DINNER

Antonio opened his restaurant less than six months ago, but one of his ideas seems to have paid off. "It's the dollar dinner," he told me, when I looked in recently. "It's a dandy but it's funny the way sales have gone."

"Should be popular at that price," I commented, for he does know food.

"I mean the weekly sales of that special dinner," explained Antonio. "One the first week, the second week eight, twenty-seven the third, and so on. Every week the cube of that week's number."

"You'll be needing more help," I suggested, trying to figure out cubes in my head.

The old rascal chuckled. "Up to the end of last week, the total sold was the square of the square of the number of girls I have now."

Maybe he made it all up for my benefit! But how many of those dinners had he sold?

33. THE STRUCTURE STILL STANDS

"What's that?" asked Ken, pointing to one of the photos.

"Something interesting," his father told him. "Older than the pyramids, but they say this is solid right through."

The boy examined the picture more closely. "See the man standing there. It must be nearly thirty feet high," he said. "Tell me about it, Dad."

"Not far off that, I guess, but there's not much to tell," his father replied. "It's a cube of exactly identical cubical blocks. Then they laid a square platform of the same blocks around its base, just one layer thick."

"Does it mean anything?" Ken asked.

His father shook his head. "Maybe it did when the ancient Kalotans built it. It seems there are twice as many blocks in the platform part as in the cube itself. That's a lot, when the blocks are around one foot each way."

For the rest of the afternoon Ken was trying to figure out how many blocks there must have been in that massive structure. How many would you say?

34.　THE DOG LEG ROUTE

"I'm driving over to Tulla today," said Bill. "Would you care to come?"

"Sure," Doug told him. "Is it far?"

"Just twenty-one miles by the straight road from here, but that's closed after last week's washout," replied his brother. "We'll have to take a longer route."

"Well, I'm in no rush." Doug laughed. "You're the boss."

Bill nodded. "There's a highway running thirty-three miles straight to Dill, then another straight highway an exact number of miles up to Tulla," he said. "But maybe it's better to cut across from the Dill highway to our closed road beyond the break by a straight old track I know. The odd thing is that we turn off an exact number of miles from here, and come to the direct road exactly the same distance along the track, and again the same distance to drive on the Tulla road."

How long would that dog leg route be?

35.　THE ELECTION

Lois looked up from her newspaper. "So Wilson won," she said. "But with less votes than the other two got together."

John nodded. "I saw that," he told her. "In fact I noticed quite an odd thing about the results. The votes polled by each pair make an exact cube. You know, a number multiplied by itself twice."

"I'll believe you." His wife smiled. "Why does Mattock have to forfeit his deposit?"

"A gimmick to discourage crackpots." John chuckled. "He got less than ten-per-cent of all the votes cast."

It was also a coincidence that the votes polled were the minimum for which those conditions could have applied.

How many votes did each of the three candidates get?

36. WHAT'S A NUMBER?

"A fraction's a number, isn't it?" asked Susan, as her brother entered the room.   "Betty says it isn't."

Dick considered the question.   "Well, I guess you'd call it a number," he replied.   "What's the point?"

"She bet me a quarter I wouldn't find a number whose square is fifteen more than the square of one number and also fifteen less than the square of another," declared his sister.   "And I figure I won the bet."

"It sounds unlikely to me, even with fractions."   Dick laughed. "Let's see this amazing number then."

Susan wrote her find on a piece of paper, and it most certainly complied with the exact requirements.

What is the simplest fraction (i.e. with smallest numerator and denominator) that would do this?

37. DUST UNTO DUST

"Tom was talking about old Dobson's Will," said John.   "I wouldn't care to be his executor."

"Eccentric as they come, I know."   Len nodded.   "But you would only have to do what he said."

John laughed.   "Exactly! He stipulated a grave-stone cut as a square-base pyramid, with all its edges and its vertical height whole numbers of feet."

"Golly!" exclaimed Len.   "I guess that's possible."

"Maybe, but listen.   He also insisted that the precise center of his casket should lie exactly nine feet from all five corners of the stone."

Pity the poor executor!   What would the dimensions of the stone be?

38. WHERE THE GRASS GREW GREEN

"I wish I could keep my grass green like yours," said Bob, leaning over the low fence as his neighbor stopped the mower. "But why the new flower beds, all triangular and all different?"

"Just an idea of mine." Professor Brayne chuckled. "The distance around each bed in feet is the same as its area in square feet, and the sides are all whole numbers of feet."

Bob pondered this a moment. "I guess that took some figuring out. How many more triangles do you plan to have just like that, still all different?"

The professor smiled. "I can't have any more," he replied. "You can see for yourself."

There was obviously plenty of space for many more on that great expanse of lawn, but Bob wasn't arguing.

How many of those special flower beds were there, and what were their dimensions?

39. THE EXCURSION

Ben pulled his bus into the curb as I was passing his office. The door slammed open and out scrambled the passengers. "How's our independent operator?" I asked, when he had finally extricated himself from a little old lady who seemed to find the step down difficult.

"Fine," he said, grinning happily. "An excursion over to Brent, and I collected just one cent over fifty bucks in fares."

"So much?" I exclaimed. "You're an old robber."

But Ben shook his head. "You saw only part of them. There was that one woman and a bunch of men and as many kids as the men, but half of the kids and half the men paid single and stayed at Brent."

That was different. "What did you charge, then?" I asked him.

"Kids paid half the adult fare for the same distance," replied Ben, "and the single fare for an adult was the same as a kid's return fare. That's taking any fraction of a cent as a full cent."

So he had not really overcharged.

What was the adult fare to Brent and back?

40. SHARING THE PROFITS

John had told them the amount of the net profit for the year:
"the exact figure," he'd said, "to the exact cent." "So now," he
went on, "I have worked out what seems a fair distribution."

He picked up his notes, with a smile for his partners who were
all there around the big table, and outlined his idea. Those net
profits would be divided by the number of partners and this, after
deduction of ten dollars, would be John's own share. "Then," he
continued, "we divide the balance by our total number and, after
deducting ten dollars, that will be Bruce's share."

There being no comment, he went on. "And then we'll follow
the same routine for each of the rest of you, in order of age. That
will leave a final balance to be divided exactly equally amongst our
sales and office staff, so that they will each get nearly $100 bonus."

"We ourselves share only $5,436.07 that way," remarked Frank.
"But I guess it isn't their fault we had a bad year."

Nobody questioned the absurd accuracy of the scheme: they all
knew the old man's foibles. So how many employees would share
that final balance?

GUIDE TO ANSWERS AND SOLUTIONS

Answers and detailed solutions, which follow now, have been arranged in the order of appearance of the items to which they refer:

ANSWERS FOR CHAPTER 7

(1) One cuts the cake into what he considers two equal portions, and then the other has first choice.

(2) Ahmed takes out what he thinks is one third of the raisins. Kemal returns some raisins to the sack as he thinks appropriate, if he considers Ahmed has taken too many: if, however, he thinks Ahmed has taken an exact third or less, he makes no adjustment. Ali then further reduces the share that Ahmed took out, or leaves it untouched, subject to the same rule. Whichever one of the three has made the last adjustment (possibly Ahmed, if the other two were satisfied) now keeps that share which he considers to be exactly a third. The other two then divide what remains in the sack by the method outlined for the two-person division of a piece of cake.

ANSWERS FOR CHAPTER 8

How Many Rectangles? In a regular chessboard there are 1296 rectangles altogether, including the 8 by 8 outer square. In the general case, a board N by N contains $(N^2 + N)^2/4$ rectangles.

ANSWERS FOR CHAPTER 9

(i) "But we get wet, lout": BUT $\times$ WE $= 125 \times 37$.

1. SALES $= 10921$.
2. DO) FLY (DO $= 29$) 841 (29.
3. SET $+$ NET $= 842 + 142$, and A LURE $= 6\ 5904$.
4. TRIED DRIVE RIVET is $17465 + 57496 = 74961$.
5. FUN IN FACT: $204 \times 14 = 2856$.
6. THE SEVEN SEVEN TEASER: $127 + 87376 + 87376 = 174879$.
7. 9 books @ 37¢, 7 pencils @ 7¢, 2 pens @ \$1.39: \$6.60.
8. IS) THAT (IT $= 56$) 2912 (52.
9. EVE) TRUSTS (ART $= 414$) 207828 (502.

10. POSH CHOP SHOP: $9758 - 3879 = 5879$.
11. LAD) TOOK (NO $= 219$) 7446 (34.
12. ICY ROAD, CAR SKIDS: $628 + 9754 + 259 = 10641$.
13. CAN) MARL (IT $= 502$) 9036 (18.
14. XMAS MAIL EARLY PLEASE is $3784 + 7860 + 98205 = 109849$.
15. SEND MORE MONEY is $9567 + 1085 = 10652$.
    HE SENT HE SENT THE TEN THEN is

$$10 + 2035 + 10 + 2035 + 510 + 503 = 5103.$$

16. EERY $\times$ OWL $= 3367 \times 198$.
17. MAN) DIZZY (MAN $= 247$) 61009 (247.
18. THREE THREE ONE SEVEN is $23577 + 23577 + 817 = 47971$.
19. $775 \times 33 = 25575$.
20. GLOW $= 4560$, FOOT $= 2661$, YARD $= 7983$, TRAY $= 1897$.
21. FOUR ONE THREE THREE ELEVEN is

$$9824 + 871 + 60411 + 60411 = 131517.$$

22. TEN TEN NINE EIGHT THREE FORTY is

$$718 + 718 + 8281 + 12347 + 74011 = 96075.$$

23. 124) 12128321 (97809.
24. SPLASH + SCOTCH = POETIC is $125013 + 169863 = 294876$,
    TOP $-$ ICE is $892 - 764 = 128 = 4^3 + 4^3$.
25. 625) 631938 (1011.1008.

SOLUTIONS FOR CHAPTER 9

No general rules can be given for the solution of alphametics and cryptarithms, but in solving such puzzles we use a few special devices that must be mentioned as an introduction to what follows.

In many solutions there will be the concept that might be expressed in words as: "seven multiplied by four ends in eight." To save writing and space it is convenient to show this as "$7 \times 4 \to 8$." Similarly, "$23 \times 32 \to 36$" will be understood.

It is often necessary to build up a tabulation from the *possible* values for one letter, thereafter tabulating the corresponding values for one or more other letters. In this process some values will be obviously unacceptable, because they entail duplication or for other reasons. Where this occurs it is convenient to strike out an unacceptable value, and such strikings out will be seen in these solutions.

If the **letter O** appears in an alphametic, it may be advisable to write the word **zero** where zero appears in the solution. This may avoid possible confusion.

If N be any even digit, $N \times 6 \to N$. If N be any odd digit, $N \times 5 \to 5$. These are useful facts.

Any square number must end in 0, 1, 4, 6, or 9. Since the letter values must be different, then, $N \times N \to M$ implies that $M = 1$, or 4, or 6, or 9, with corresponding values for N.

Finally, although obvious because all such puzzles must conform to the normal practice in the corresponding numerical calculations, the first or initial digit of a written number cannot be zero.

The solutions, only outlined in most cases, now follow.

(1)
```
      L  O  S  E
      S  E  A  L
      S  A  L  E  S
```
$S = 1$, so $A = $ zero, $L = 9$.
From $L + S$, $E + L = 11$.
So $E = 2$. Hence $O = 7$.

(2)
```
D  O  )  F  L  Y  (  D  O
         I  F
         D  R  Y
         D  R  Y
         -  -  -
```
$D \leq 3, D > 1$.
If $D = 3$, then $I = 9$ which is impossible.
So $D = 2$.
From $DO \times O$, $O > 7$.
Hence $DO = 28$ or $\quad$ 29
with $\quad$ FLY $= 784 \qquad 841$
$O \neq L$, so $DO = 29$, $FLY = 841$.

(3)
```
      S  E  T
      N  E  T
      U  S  E
            A
   L  U  R  E
```
$ET - ET$ would imply $S = E = $ zero.
So $SET + NET = USE$, and E is even.
Then, as multiplier, $A = 6$.

$T =$	1	2	4	5	7	9
with $E =$	2	4	8	0	4	8
and $S =$	4	8	$\cancel{6}$	1	9	7
with $R =$	5	0	—	$\cancel{6}$	$\cancel{6}$	$\cancel{6}$
and $U =$	—	9				
$N =$	—	1				

(4) T R I E D   If subtraction, D < T. Then, from
    D R I V E   E − V, we have V = zero. This leads
    R I V E T   to I = zero, impossible.

So this is addition and, from E + V, V = zero or 9.

Say V = zero.   Then I = 5, making R = 2 or 7. But R > 2,
                 so R = 7, and T + D = 6. From D + E,
                 T > D, so D = 2, T = 4, and E = 2, which
                 is impossible.

So V = 9.   Then I = 4, making R = 7. T + D = 6,
                 But T < D, so T = 1, D = 5, E = 6.

(5)   F U N   I = 1. N > 1 so F < 5. F × N < 10.

	I	N		$N =$	2	3	4
*	*	*		$T =$	4	9	6
*	*	*	$F × N < 10$, so	$F =$	3	2	2
F	A	C	T	$IN =$	12	13	14
			Maximum	$FAC =$	398	287	298
	whence, without duplication,			$FUN =$	302	203	204
			making	$FAC =$	36̸2̸	26̸3̸	285

So FUN = 204, FACT = 2856.

(6)   T H E   T = 1, and S > 4.
        S E V E N From HEE there must be "carry 1,"
        S E V E N so T + 2V + 1 → S,
     T E A S E R i.e.,   2V + 2 → S, so S is even.

| | $S =$ | 6 | | | | 8 | | |
	$E =$	2				7		
	$2V = 4$			14		6		16
	$V = —$			7		3		—
	$H + E = —$	10	9	8		10	9	8	—
with	$H = —$	8	7̸	6̸		3̸	2	1̸	—
and	$2N =$	$R − 2$				$R + 3$		
so	$R =$	4				5	9	
with	$N =$	1̸				4	6	
and	$A =$	—				4̸	4	

So we have TEASER = 174879.

(7) L 40082   @   NC ¢ .. .. N.NN
    C 1637592   @   C ¢ .. ..   .BL
    S 1632   @ $P.NL   .. ..   S.CK
                                                         $E.EO

NNN is a multiple of 37, and also of NC. If C is even, N must be even, so NC $\neq$ 74. Hence NC = 37, whence BL = 49. From S.CK, P = 1, hence S < 4. So S = 2, K = 8. Summing we get E = 6, O = zero. Only 5 remains unused, so I = 5.

(8) I S) T H A T (I T   From S $\times$ T, S = 6 or
          T R Y   T = 5. If T = 5, IS < 20,
           ? ? ?   but I $\neq$ 1. So S = 6,
           ? ? ?   and T is even. Then,
           — — —   to get TRY, I > 4.

                       IS = 56   76   86   96
     with TRY = 280   532   688   864

Here, TRY = 280 is the only acceptable value, so THAT = 56 $\times$ 52 = 2912.

(9) E V E) T R U S T S (A R T   R = zero
                * * * *   E < 5, T < 5,
                     * * *   E > 1, T > 1,
                     * * *   E $\times$ T < 10
                     — — —   so E = 2 2 3 4
                            with T = 4 3 2 2

But 1st digit of product **** must be less than E, and must also be either equal to or 1 less than T. Hence T = 2, with E = 3 or 4.

Tabulate:

	E =	3	4
	T =	2	2
	S =	6	8

	*** starts with	6 or 7	8 or 9
	making **** → 0	9	0 9
	whence A = ∅	3	5 —

Then, TS = 28, so *** = 828, hence EVE = 414.

(10) P O S H   If addition, H = S = zero.
<u>C H O P</u>   So this is subtraction.
S H O P   From P − C, P > 2, also P > S.

Only part of the necessary tabulation is given. If completed, that will yield the acceptable solution.

	P = 3	4	etc.
	H = 6	8	etc.
from O − H,	O = 2	6 or 7	etc.
from S − O,	S = 4	2 —	etc.
which would give	O = —	7 —	etc.
with	C = —	— —	etc.

(11) L A D) T O O K (N O   O > 1, N > 1, so
       <u>K I T</u>   L < 3, hence L = 1
        * * *   or 2. K > 2, T > 3,
        <u>* * *</u>   T ≠ 5, K ≠ 5, so
        — — —   T ≠ 6, N < K.
                          Re possible factors,
                          T ≠ 9.

Tabulate for T = 8, 7, 4, completing the scheme shown.

	T =	8		
	K =	7		
from D × N, and N < K,	N = 2	3	4	6
with	L = 1	2	1	1
from D × N,	D = —	6	2	3
from D × O,	O = —	—	—	9
so TOOK = —		—	—	8997
dividing by NO, LAD = —		—	—	—

(12)   I   C   Y   C   C   C $S = 1$, $K = $ zero, $R = 9$

      R   O   A   D   N   O   So $D + Y = 12$, $Y \neq 6$

        C   A   R   C   A   R   Also $2A + C \rightarrow D - 2$

   S   K   I   D   S

Complete tabulation for $Y = 4, 5, 7, 8$, leading to values of CCC, CAR, and O. Thence value of N, and the remaining digit will give I.

$$Y = 4$$
$$D = 8$$
$$2A + C \rightarrow 6$$

$C = 2$	6
$A = 7$	5
'carry' from CAA .. 1	1
$O = 7$	3
$CCC = -$	666
$NO = -$	N3
$CAR = -$	659
$N = -$	—

(13) C   A   N   )   M   A   R   L   (I   T $I = 1$, $A = 9$ or zero

           C   A   N   so $R \rightarrow N + 1$.

         S   A   I   L   If $A = 9$, $N + 1 =$

         *   *   *   * $R + 10$,   whence

         −   −   −   − $R = $ zero, $N = 9$.

                               So, $A = $ zero, $R = N + 1$.

Now, $AN \times T \rightarrow IL$, and $A = $ zero and $I = 1$, so $N \times T < 20$. $S \neq 1$, and $T > S$, so $T > 2$, hence $N < 7$. $M = C + S$. $S > 1$, $S < C$, so $C > 2$, hence $M > 4$. Also, $N \times T \geq 10$, and $T \neq 5$.

Complete scheme of tabulation as shown, taking into account all possible values for T within limitations.

$N = $	2 etc.
$R = $	3
$AN = $	02
$T = $	7
$IL = $	14
$IT = $	17
$ARL = $	034

Re division by IT, $M = -$

(14)

```
      X   M   A   S
          M   A   I   L
      E   A   R   L   Y
  ─────────────────────
  P   L   E   A   S   E
```

$P = 1$, $L = $ zero, $E = 8$ or 9. If $E = 8$, then $X + M + A = E + 19$ so $X + M + A = 27$, impossible. Hence, $E = 9$.

Then, $S + Y = 9$. Also, $X + M + A = 18$, so $A > 2$, $M > 2$. Complete tabulation, based on $S + Y = 9$, as indicated.

S =	2		etc.
Y =	7		etc.

A + I →	2	
A =	8	4
I =	4	8

M + R =	9		9	
R = 3	6	3	6
M = 6	3	6	3

| X = | 4̶ | 7 | 8̶ | 11 |

(15)

```
          H   E
  S   E   N   T
          H   E
  S   E   N   T
      T   H   E
      T   E   N
  ──────────────
  T   H   E   N
```

$3E + 2T →$ zero, so E is even. $T \geq 2S$, so $T \geq 2$. Then we have: $E = 0 \quad 2 \quad 4 \quad 8$ with $T = 5 \quad 7 \quad 9 \quad 3$ Complete tabulation, noting the 'carry' to the next column wherever appropriate.

E =	8	etc.
T =	3	etc.

'carry' =	3
3H + 2N →	7

H =	1			5		7	9	
N =	2	7	1	6		—	0	5

'carry' =	1	2	2	3	—	3	4
giving H =	3̶	4̶	4̶	5	—	5̶	6̶

'carry'	—	—	—	2	—	—	—
2S =	—	—	—	1	—	—	—

(16)

```
        E  E  R  Y
           O  W  L
     ─────────────────
     *  *  *  *  *
  *  *  *  *  *
  *  *  *  *
  ─────────────────
  R  R  R  R  R  R
```

$Y \times L \rightarrow R$, so $R \neq 9$, $L \neq 9$. If $E = 1$, 1st product could not exceed 1176×8, i.e., 9408, so $E > 1$. $E < R$, so $E < 8$, and $R > 2$. $Y \neq 5$, $L \neq 5$, $R \neq 5$. Hence $R = 3, 4, 6, 7,$ or 8.

If EERY is a multiple of 11, $R = Y$, impossible. But RRRRRR is a multiple of 11, so OWL is also a multiple of 11, whence $L + O = W$ or $W + 11$. $E > 1$, so $O < 5$. If $O = 4$, $R \geq 9$ (i.e., $2 \cdot 4 + 1$). Hence, $O = 1, 2,$ or 3. Also $W > 0$, and $L > 0$.

Tabulate:

$O =$	1						2				3	
$W =$	3	4	5	7	8	9	5	6	8	9	7	9
$L =$	2	3	4	6	7	8	3	4	6	7	4	6

Now list these values for OWL with their factors:

$132 = 11 \cdot 3 \cdot 4$	$187 = 11 \cdot 17$	$286 = 11 \cdot 13 \cdot 2$
$143 = 11 \cdot 13$	$198 = 11 \cdot 3 \cdot 6$	$297 = 11 \cdot 3 \cdot 9$
$154 = 11 \cdot 7 \cdot 2$	$253 = 11 \cdot 23$	$374 = 11 \cdot 34$
$176 = 11 \cdot 16$	$264 = 11 \cdot 3 \cdot 8$	$396 = 11 \cdot 3 \cdot 12$

$RRRRRR = 3 \cdot 7 \cdot 11 \cdot 13 \cdot 37 \times R$, so all the factors of OWL must be amongst these (and $R < 9$). So, as $R < 9$, OWL $\neq$ 176, 187, 253, 297, 374, or 396. Individual study of the remaining 6 values will eliminate all except OWL $= 198$, with EERY $= 3367$.

(17) M A N) D I Z Z Y (M A N
```
             *  *  *
           ─────────────
           *  *  D  *        Y = 1  4  4  6  9  9
           *  *  R           N = 9  2  8  4  3  7
           ──────────        If M = 3, D > 9, so
           *  *  *  Y        M = 1 or 2.  N > M,
           *  *  *  *        and N > A.
           ──────────
           -  -  -  -
```

Say M = 1.   Then DIZZY > 20333, N > A, so MAN > 146
              MAN × A < 1000, so MAN < 167 but N > A,
              hence MA < 16.

MA =		14			15	
N =	7	8	9	7	8	9
Y =	9	4	1	9	4	1
R =	8	—	—	5	0	—
MAN =	147	—	—	—	158	—
DIZZY =	21609	—	—	—	24969	—

Neither value is acceptable. So, M = 2, and similar procedure will yield the required solution.

(18)

T	H	R	E	E
T	H	R	E	E
		O	N	E
S	E	V	E	N

The solution of this is interesting because it exemplifies two useful devices which simplify the working. The puzzle could be solved by more usual methods, of course.

We have $3E = $ N   or $N + 10$ or $N + 20$

with $E + N = $ 10   9   8

whence $3E = 10 - E$ $19 - E$ $28 - E$

so $E = $ —   —   7

hence, $E = 7$, $N = 1$, and from $H + H$ we have $H = 3$ or 8. Before proceeding, we consider divisibility by 7. To test any number for divisibility by 7, we mark off the digits "from the right" in threes. The condition for such divisibility is that the *difference* between the sum of the odd trios (1st., 3rd., etc.) and the sum of the even trios (2nd., 4th., etc.) "from the right" must be divisible by 7. For example, 324294558 must be divisible by 7, because $(558 + 324) - (294) = 588 = 84 \cdot 7$.

Now tabulate:

H =	3			8	
T =	2	4	2	3	4
S =	4	8	5	7	9
SEVEN =	47V71	87V71	57V71	—	97V71
By the rule, V =	2 9	1 8	0 7	—	6

Thence, corresponding values for $2R + O$ can be noted, so leading to the required solution.

(19)

```
        X  Y  C
           A  B
      *  *  P  D
   *  *  *  E
   *  *  *  R  D
```

We set this out as shown in order to simplify identification in the solution, but without regard to possible values of the digits in question. We are restricted to 2, 3, 5, 7.

$$2 \times 2 = 4, \quad 2 \times 3 = 6, \quad 2 \times 5 \to 0, \quad 2 \times 7 \to 4,$$

so $A \neq 2$, $B \neq 2$, $C \neq 2$.

Testing 3, 5, and 7 in this way, we find that $D = 5$, $E = 5$, and A, B, C must conform with:

$$A, B = 3 \qquad 5 \qquad 7$$
$$\text{with } C = 5 \quad 3, 5, 7 \quad 5$$

So the possible trios can be tabulated:

AB =	33	35	37	55	55	55	57	77	75	73	53
C =	5	5	5	3	5	7	5	5	5	5	5

Each trio must now be tested, inevitably a laborious process! In this testing, it should be noted that $P = 2$ or 7, with correspondingly $R = 7$ or 2, because $E = 5$. Here, for example, we test AB = 73, C = 5.

If   AB = 73,   XYC > 2221/3,   so XYC ≥ 755

Test by multiplication, stopping if an unacceptable digit appears.

```
    755          775
     73           73
    ₲5         2325
               ₳25
```

Similarly we would test the other 10 possible values for AB, and so discover the unique solution.

(20)

F	O	O	T		T	R	A	Y		$T + F < 10$, so $T < 9$.
F	O	O	T		F	O	O	D		$3F < 10$,
F	O	O	T		G	L	O	W		so $F = 1, 2,$ or 3.
Y	A	R	D		(addition)					From $A + O$, A = zero
										or 9.

Complete tabulation, on scheme shown:

		A =	zero
From OOO........	O =	3	6
with	R =	1	—
'carry' from 3T =		2	—
In GLOW........	L =	4	—
From F < 4........	F =	2	—
making	Y =	7	—
From 3T, and 'carry'	T =	—	—
and	D =	—	—

(21)

	F	O	U	R
		O	N	E
T	H	R	E	E
T	H	R	E	E
E	L	E	V	E	N

Because FOUR is a multiple of 4, UR must be a multiple of 4. $E = 1$ or 2. R is even.

Say $E = 2$. Then $T = 9$, $L = $ zero. $N \to R + 6$, so $R = 8$, $N = 4$, whence $U = 3$. This makes $UR = 38$.

So $\quad E = 1$, $\quad N \to R + 3$, $\quad U \to 9 - (N + \text{'carry'})$,

$\qquad V \to 2R + 2O + \text{'carry'}$.

Complete the tabulation for $R = $ zero, 2, 4, 6, 8.

$$R = \qquad 0$$
$$N = \qquad 3$$
$$U = \qquad 6$$
$$UR = \qquad 60$$

$O = 2$	4	
$V = 5$	9	
'carry' $= 0$	0	
$F + 2H \to 1$	1	
$F = -$	7	5
$H = -$	2	8
'carry' $= -$	1	2
$T = -$	—	—
$L = -$	—	—

(22)

```
        T  E  N
        T  E  N
     N  I  N  E
  E  I  G  H  T
  T  H  R  E  E
  F  O  R  T  Y
```

NINE is a square, $E \neq$ zero, $E \neq 9$, so $E = 1, 4, 5,$ or 6. By quick trial (or a table of squares) we find that NINE must be one of:

2025, 3136, 6561, 8281.

Each of these possibilities must be examined separately.

For example, we take NINE = 8281, showing the method for some values. The same procedure should be applied for $T = 7$, and then for the other three values of NINE.

$$N = 8, \quad I = 2, \quad E = 1, \quad N + I + H \geq 10, \quad T + 8 \rightarrow Y.$$

T = 5	6	
Y = 3	4	
H = 2	3	
Re N + I + H + 'carry', ... F = —	8	9
'carry' from EENHE = —	—	1
I + G → —	—	7
so G = —	—	5
'carry' = —	—	2
N + I + H + 'carry' = —	—	15

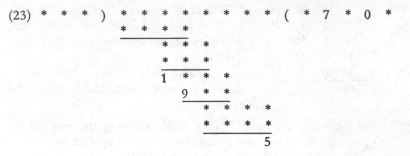

Three of the unknown values are obvious by inspection, and these have been filled in accordingly. The 2nd sub-product, * *-*, must start with 7 or 8, so divisor $< 900/7$, i.e. divisor < 129. From 3rd sub-product, 9 * *, 3rd digit of quotient > 7, hence the quotient must be 97809. Then,

$$\text{divisor} < 1000/8, \quad \text{i.e. divisor} < 125.$$

Say divisor $= x$, and dividend $= 1000y + z$, with $z < 1000$.

We have $\qquad 1000y + z = 97809x + 5$
and also $\qquad\quad y - 97x \geq 100$
i.e., $\qquad\qquad 1000y = 97809x - z + 5$
and also $\qquad\quad 1000y \geq 97000x + 100000$
hence $\quad 97809x - z + 5 \geq 97000x + 100000$
so $\qquad\qquad\quad 809x \geq 99995 + z$
whence $\qquad\qquad\quad x \geq 124$

but $x < 125$, so $x = 124$. The divisor being 124 and the quotient 97809, the dividend is obtained.

(24) S P L A S H   TOP − ICE = the sum of 2 cubes.
      S C O T C H   Here, T > I, so I ≠ 9, T ≠ zero.
      P O E T I C   A = 9 or zero,   S < 5,
                             P = 2S or 2S + 1.

The solution involves tabulation of possible S values, with corresponding values for other letters.   Where this yields complete allocation of all digits we check for TOP − ICE = the sum of 2 cubes.   For example, here we consider S = 3, P = 7.

	H = 1	4	8	9
	C = 2	8	6	8
	I = 5	1	0	2
	A = 0	9	9	$\emptyset$
From P + C....	O = —	5	6	4	—
3 digits remain:	L = —	0	5	—	—
	E = —	6	2	—	—
Re P + C, check,	O = —	5	6	—	—
Final digit gives	T = —	2	$\emptyset$	—	—

But here, TOP − ICE = 257 − 186 = 71, which is not the sum of two cubes.

(25) * * a) * * * * * * * (* 0 * * . b 0 0 c
          * * *
          ‾‾‾‾‾‾‾
            * * *
            * * *
            ‾‾‾‾‾‾‾
              * * *
              * * *
              ‾‾‾‾‾‾‾
                * * 0
                * * d
                ‾‾‾‾‾‾‾
                e 0 0 0
                e 0 0 0
                ‾ ‾ ‾ ‾

Some of the key digits have been labelled (i.e., a, b, c, d, e) in order to simplify identification. Obvious zeros have been written in.

To obtain the final sub-product, e 0 0 0, we must have $c = 5$ with a = zero, or else $a = 5$.

But, if a = zero, then d = zero, and e = zero, which would be absurd.

Hence, $a = 5$ and c is even. Then the divisor must end with 25 or 75, so $d = 5$, making $e = 5$. Hence, as a factor of 5000, divisor is 625. Inspection of the sub-products, for divisor > 500, shows immediately that the quotient must be 1011.1008.

ANSWERS FOR CHAPTER 11

1. Tether approximately 68 yards long.
2. 4 boys, 15 blue and 6 green marbles.
3. 4 girls shared $170.00.
4. Ages of the "girls": 7, 7, and 50 years.
5. Ben was 4 blocks south and 4 blocks east from Bert.
6. Joe took Steve's hat, Ron's coat: Steve took Ron's hat, Doug's coat.
7. Hank started with 40¢.
8. Ages were: John 44, Ted 39, Ben 35 years.
9. Johnnie earned $47.41 his first week.
10. Bob averaged 72 m.p.h. for the final seventh of the drive.
11. 61 soldiers.
12. 24 days when it rained.

13. 24 blouses @ 48¢, 23 @ 47¢, 22 @ 46¢.

14. Ted's mother 35 years old.

15. 216 coins: Ken had 54 dimes, Pete 162 nickels.

16. Pool would be 20 feet long, 13 feet wide.

17. The second visit was on a Wednesday.

18. Kiko had 52 kuks, 23 dimes; Peter 43 kuks, 7 dimes.

19. Wife's age 34, son's age 13 years.

20. 2 frames @ $2.10, 5 @ $3.30, 3 @ $4.62, 4 @ $7.70, 3 @ $11.55.

21.   "You get that rich refrain, we think,
           If three and forty goblets clink."

22. 324 tiles, each 3 inches square.

23. 14 games, 8 players.

24. 4 cakes @ 12¢, 6 @ 14¢, 4 @ 17¢.

25. Manitoba: 198 (including 12 foremen); Ontario: 214 (including 12 foremen); Quebec: 188 (including 12 foremen).

26. Paul was reading page 342.

27. Peter had 3125 soldiers.

28. The diameters were 17, 14, and 7 kebals.

29. Jake bought 4 geese, 8 turkeys, 8 ducks.

30. Charlie took Andy's hat.

31. She owed $26.73, from solution of the "indeterminate equation" $98x - 199y = 20$.

32. 1296 "dollar dinners" had been sold.

33. 41,472 blocks altogether in the complete structure.

34. The dog leg route was 33 miles.

35. Wilson, 6654 votes; 2nd candidate, 5513; Mattock, 1346.

36. The simplest fraction would be 17/4.

37. Base edge 8 feet, slant edge 6 feet, vertical height 2 feet.

38. 5 flower beds: 6, 8, 10; 5, 12, 13; 9, 10, 17; 7, 15, 20; 6, 25, 29 feet.

39. Adult return fare was $1.94.

40. 31 employees would share $2869.36.

SOLUTIONS FOR CHAPTER 11

2. A GAME OF MARBLES

The possible different selections, observing that each boy had to have at least 1 green marble, were:

R	9	8	7	6	7	6	5
B	2	3	4	5	3	4	4
G	1	1	1	1	2	2	3

Pete had more than 1 green, but did not have 4 blue, so he had 7 red, 3 blue, 2 green.

The remaining boys had 19 red, involving the selections

R	7 OR	6 OR	5 and	9	8	6
B	4	4	4	2	3	5
G	1	2	2	1	1	1

Only one arrangement will yield 19 red:

R	5	8	6
B	4	3	5
G	3	1	1

So there were 4 boys, with 15 blue and 7 green marbles.

3. THE BONUS

Men: 3 points per year, total x years .... $3x$ points
Girls: 2 points per year, total $(23 - x)$ years .... $(46 - 2x)$ points. So points totalled $(x + 46)$, necessarily a factor of 26000. Hence, $k(x + 46) = 2600 = 2^4 \cdot 5^3 \cdot 13$, k being an integer. There was at least one girl, so $x \leq 21$, whence $k \geq 26000/67$, i.e., $k > 388$.

Also, $(x + 46) \geq 46$,   so $k < 565$.

The only sets of factors of 26000 that give values of k within the acceptable limits lead to:

$$
\begin{array}{rccc}
k = & 400 & 500 & 520 \\
x + 46 = & 65 & 52 & 50 \\
x = & 19 & 6 & 4
\end{array}
$$

Years of service were 2, 3, 5, 6, and 7 respectively, and no combination of these will total 4 years or 19 years. So we have $x = 6$, with $k = 500$ (i.e., $5.00).
Also, 6 is uniquely represented from the years of service by the single number 6. Hence there was only 1 man, and he had 6 years of service.
Then the other four were girls, with 2, 3, 5, and 7 years of service. Hence the allocation must have been:

Man....... 18 points @ $5.00...... $ 90
4 girls...... 34 points @ $5.00...... $170

4. THE JOY-RIDE

Grouping the factors of 2450, within reasonable limits of ages, we have "possibles":

2	2	5	5	5	7	7	7
25	35	7	10	14	7	10	14
49	35	70	49	35	50	35	25
76	72	82	64	54	64	52	46

The Professor said: ". . . that still leaves their ages in doubt." So his age must have been 32, 64 being the only sum that appears twice.

If Bill's age, which the Professor knew, had been 48 years or less, there would still have been no certainty after his final statement. So Bill must have been 49 years old, the ages of the three "girls" being 7, 7, and 50 years.

5. WOULD THEY MEET?

Going from intersection A to intersection B, m blocks "one way" and n blocks "the other way," the total number of different routes is:

$$\frac{(n + 1)(n + 2) \ldots (n + m)}{m!}.$$

In this case we have $m = 4$,

hence $\dfrac{(n + 1)(n + 2)(n + 3)(n + 4)}{24} = 70,$

whence $n = 4$.

So Ben was 4 blocks south and 4 blocks east from Bert.

6. AFTER THE PARTY

This can be solved neatly by the Boolean approach (see Chapter 5).

Assign code letters as follows:

Man's Name	Himself	His hat	His coat
Ron	R	r	1
Doug	D	d	2
Steve	S	s	3
Joe	J	j	4

Man who took Joe's hat took Steve's coat:

so
$$Rj3 + Dj3 = 1 \dots\dots\dots\dots\dots\dots\dots (1)$$

Hat taken by Steve belonged to man whose coat was taken by Joe:

so $J1 \cdot Sr + J2 \cdot Sd + J3 \cdot Ss = 1$,   but $Ss = 0$,
hence $J1 \cdot Sr + J2 \cdot Sd = 1 \dots\dots\dots\dots\dots\dots (2)$

But, Ron took Doug's hat, so $Rd = 1$, $Rj3 = 0$, hence re equation (1), $Dj3 = 1$. Also $Sd = 0$, so $J2 \cdot Sd = 0$, whence re equation (2), $J1 \cdot Sr = 1$.

So, re hats we have: $Rd = Dj = Sr = 1$, hence $Js = 1$ and re coats we have: $D3 = J1 = 1$. And, because $Rd = 1$ implies $R2 = 0$ (one man's coat, another man's hat), we must have $R4 = 1$, hence $S2 = 1$.

So,

Joe took	Steve's hat,	Ron's coat
Steve took	Ron's hat,	Doug's coat
Ron took	Doug's hat,	Joe's coat
Doug took	Joe's hat,	Steve's coat

7. A MILD GAMBLE

Say A started with $5x$, B with $(75 - 5x)\cent$, A being the loser of the 1st game.

If A lost all games: $75 - 4x = \pm7$, whence $x = 17$, which is impossible.

If A lost 1st and 3rd games: after 1st game, B would have $(75 - 4x)$, which is not divisible by 4, and so is impossible.

If A lost 1st and 2nd games: $(2x + 25) - (50 + 2x) = \pm7$, whence $x = 8$.

Then, A started with $40\cent$, B with $35\cent$.
              Ben lost the 3rd game, so Ben was B.
Hence,   Hank started with $40\cent$.

8. A FRIEND IN NEED

Say ages were: John x, Ted y, Ben z years, with $x > y > z$; and x, y, z all integral.

$$z(x - y) = x(y - z) - 1, \text{ whence } (x + z)(2x - y) = 2x^2 - 1.$$

So, $(2x^2 - 1)$ has two factors, $(x + z)$ and $(2x - y)$. Obviously, neither of these can equal unity, hence $2x^2 - 1$ cannot be a prime number.

x must lie between 43 and 49. So $x = 44$ or 47, because all other values in that range make $2x^2 - 1$ prime.

Taking the factors of $2x^2 - 1$ for $x = 44$ and for $x = 47$, and observing that all ages must be adult, the unique acceptable solution follows: $x = 44, y = 39, z = 35$.

9. HIS FIRST JOB

Say he earned $x\cancel{c}$ his first week, and worked y weeks.

Then, $xy + 99y(y - 1)/2 = 40700,$

whence $y(2x + 99y - 99) = 81400 = 2^3 \cdot 5^2 \cdot 11 \cdot 37.$

Now, $x > 4000,$ so $99y^2 + 7901y < 81400,$

hence $y < 10,$

also, $x < 6000,$ so $99y^2 + 11901y > 81400,$

hence $y > 5.$

But, y must be a factor of 81400, so $y = 8$ and, substituting for y, we get $x = 4741.$

So, Johnnie earned \$47.41 his first week.

10. TO THE COTTAGE

Say total distance was x miles.

$$\text{1st part} - 15 \text{ miles @ } 40 \text{ m.p.h.} - \frac{3}{8} \text{ hours}$$

$$\text{2nd part} - \frac{ax}{9} \text{ miles @ } \quad b \text{ m.p.h.} - \frac{ax}{9b} \text{ hours}$$

$$\text{3rd part} - \frac{x}{7} \text{ miles @ } \quad c \text{ m.p.h.} - \frac{x}{7c} \text{ hours}$$

where $c > b > 40$, and a, b, c are $+ve$ integers.   Then

$$\frac{x}{7} + \frac{ax}{9} + 15 = x, \quad \text{whence } x = \frac{945}{54 - 7a}.$$

$(54 - 7a)$ must be $+ve$, so $a < 8$, and $x > 100$, so $a > 6$.   Hence,
$a = 7$, $x = 189$.   Total time was

$$\frac{3bc + 1176c + 216b}{8bc} = \frac{189}{56} = \frac{27}{8} \text{ hours}$$

so $\qquad\qquad\qquad bc + 72b + 392c = 9bc$

i.e., $\qquad\qquad\qquad bc - 9b - 49c = 0$

hence $\qquad\qquad (b - 49)(c - 9) = 441 = 3^2 \cdot 7^2$

No "crazy speeds", so we can assume $(c - 9) < 147$ and obviously
we can assume $(c - 9) > 21$.

So we have $\qquad\qquad c - 9 \ = 49 \quad \text{or} \quad 63$

with $\qquad\qquad\qquad b - 49 = \ 9 \qquad\qquad 7$

making $\qquad\qquad\qquad\qquad c = 58 \qquad\qquad 72$

and $\qquad\qquad\qquad\qquad\quad b = 58 \qquad\qquad 56$

but $c > b$, hence $\qquad\qquad c = 72, \quad b = 56.$

Average speed for the final seventh was 72 m.p.h.

12. THAT ENGLISH WEATHER

Say x days rain, y days dry, and he smoked a cigarettes each dry day (a being an integer).

Average consumption for the period was $(20x + ay)/(x + y) = 5a$,

hence $\qquad\qquad 5x(4 - a) = 4ay$, so $a < 4$.

Also, $\qquad\qquad\qquad x > y$, so $a > 2$.

Then $\qquad\qquad\qquad a = 3$, whence $5x = 12y$.

So $(x + y)$ must be a multiple of 17, "about 5 weeks". Hence,

$$x + y = 34, \quad \text{with } x = 24, \quad y = 10.$$

So there were 24 rainy days.

13. ONE CENT UP, ONE CENT DOWN

Say he bought x @ $(y + a)\cent$, $(x - y)$ @ $(y + b)\cent$, $(x + 1)$ @ $(y + c)\cent$, where a, b, c are $+1$, 0, -1 not necessarily respectively.

Then total cost was $3xy + (c - b) + x(a + b + c)$ cents

$a + b + c = 0$, so total cost was $3xy + (c - b)$ cents.

$3245 \equiv 2 \equiv -1 \pmod{3}$, hence $c - b = 2$ or -1.

So, we have	$c =$	1	or	0	or	-1
with	$b =$	-1		1		0
and	$a =$	0		-1		1
Hence	x @	y		$y - 1$		$y + 1$
	$x - 1$ @	$y - 1$		$y + 1$		y
	$x + 1$ @	$y + 1$		y		$y - 1$
3245	$=$	$3xy + 2$		$3xy - 1$		$3xy - 1$
xy	$-$	1081		1082		1082

Now, $1082 = 2 \times 541$, giving no acceptable integral x.

But, $1081 = 23 \times 47$, and we know that $3x$ was "about 72," hence $x = 23, y = 47$.

So, he bought 24 @ $48\cent$, 23 @ $47\cent$, 22 @ $46\cent$.

14. FOUR GENERATIONS

Ages, in years: Mrs Ryle $10a + b$, her father $10x + y$.
Then her grandfather's age was:

$$(10x + y)(a + 10b) - (x + 10y)(10a + b) = 99(bx - ay),$$

hence, her grandfather's age was 99, and $bx - ay = 1$.

Also, $10x + y + 10a + b = 99 - 6,$

so $10(x + a) + (y + b) = 93,$

hence $x + a = 9$ or   8,

with $y + b = 3$ 13.

If $x + a = 9$, then $a = 9 - x, \quad b = 3 - y,$

giving $3(x - 3y) = 1,$ which is impossible

so $x + a = 8$, with $a = 8 - x, \quad b = 13 - y,$

giving $13x - 8y = 1,$

with general solution $x = 8k - 3, \quad y = 13k - 5.$

x and y being integers, $x = 5, \quad y = 8,$

hence $a = 3, \quad b = 5.$

So the ages were: Mrs Ryle's grandfather.... 99 years
                 Her father.............. 58
                 Herself................. 35

15. IN THE HOBBY SHOP

Ken had x coins each $a¢$, Pete had $3x$ coins each $b¢$, then $4x = y^3$, where y was number of coins in each pile.

Also, $x(a + 3b) = 1350,$

hence $y^3(a + 3b) = 5400 = 2^3 \cdot 3^3 \cdot 25,$ and y is even.

Then $y = 2$,   with $a + 3b = 675,$ or $y = 6$, $a + 3b = 25,$

but a and b are two different numbers in the set 1, 5, 10, 25, 50 hence $(a + 3b)$ cannot exceed 175.   So

$y = 6$, $a + 3b = 25$,   making $a = 10$, $b = 5$,   and $x = 54$.

Ken had 54 dimes, Pete had 162 nickels.

16. ONE UP ON THE JONESES

Say dimensions were x feet by y feet, the cost $\$C$, with x, y, C being integers.

Then   area $= (Cx - 35x^2)/27$ square feet, an integer.

For maximum area, $x = C/70$, $y = C/54$,   and $C \leq 999$.

So,   maximum area $= C^2/3780 = 999^2/3780 = 264$ plus a fraction.

$$264 = 2^3 \cdot 3 \cdot 11, \quad 263 = \text{prime}, \quad 262 = 2 \cdot 131,$$
$$261 = 3^2 \cdot 29, \quad 260 = 2^2 \cdot 5 \cdot 13, \quad \text{etc.}$$

For maximum area we must have $x/y = 27/35$ approximately. With factors of 264, the nearest would be $x = 12$, $y = 22$, i.e. $x/y = 6/11$, making $C = 1014$ which is unacceptable.   With factors of 261, the nearest would be $x = 9$, $y = 29$, a poor approximation, making $C = 1098$, also unacceptable.   With factors of 260, the nearest would be $x = 13$, $y = 20$, a better approximation, making $C = 995$.

So the dimensions would be 20 feet long by 13 feet wide.

17. THE CLOCK

The clock loses 7 minutes per hour, so after x hours it has lost $7x$ minutes.

Say it shows correct time after x hours.

Then, it has lost

$$\{(720 - 3) + (12 \times 60 \times y)\} \text{ minutes, where } y \text{ is an integer.}$$

Hence $7x = 717 + 720y$

with solution: $\left.\begin{array}{l} x = 720k + 411 \\ y = 7k + 3 \end{array}\right\}$ k being some integer

For the 1st time the clock was correct, after first visit, $k = 0$, with $x = 411$, $y = 3$.   This is 17 days and 3 hours after that Sunday.

Thereafter, the clock would be correct at intervals of 720 hours, i.e., 30 days.

The 2nd time the clock could be correct, after first visit, would therefore have been on the 47th day after it.   That could not have been before well into the following month.

Hence, the second visit must have been on the 17th day after the Sunday, so it was on a Wednesday.

18. ONLY SMALL CHANGE

Say they had:

$$\text{Kiko} - x \text{ kuks, } z \text{ dimes}$$
$$\text{Peter} - y \text{ kuks, } \frac{2x - 3y + 2z}{3} \text{ dimes}$$

applying full exchange rate, $68x - 27y + 95z = 4560$
where x, y, z are integers.

Dividing through by 27, $\dfrac{13x + 13z - 3}{27}$ is an integer,

hence $\dfrac{x + z + 6}{27}$ is an integer, say k.

Then, $x = 27k - z - 6, y = 68k + z - 184$.
But, z dimes was worth $19z/10$ kuks, and Peter's y kuks were insufficient to pay in full. So $10y < 19z$.
Hence, $680k + 10z - 1840 < 19z$, so $680k < 9z + 1840$.
But Kiko had less than 24 dimes, so $z < 24$, whence $680k < 2056$,
hence $k < 4$.
But, from expression for y, as $z < 24$, $k > 2$.
Then k, being an integer, must be 3.
So the solution becomes: $x = 75 - z, y = 20 + z$
But $10y < 19z$, so $200 + 10z < 19z$, whence $z > 22$.
So $x = 52, y = 43$, with $z = 23$.
Kiko had 52 kuks, 23 dimes: Peter had 43 kuks, 7 dimes.

19. THE JOINT BIRTHDAY

Say the ages were: mother x, son y years, x and y being integers.
We derive the equation $X^2 - 5Y^2 = -4$,

$$\text{where} \quad \left. \begin{array}{l} 2x - 3y = X \\ y = Y \end{array} \right\}$$

This has two distinct "families" of solutions, based on $X = 1$,
$Y = 1$, and $X = 4$, $Y = 2$ (see Chapter 6). Thence, the successive
pairs of integral values are:

	$X = 1$	4	11	29	76	etc.
	$Y = 1$	2	5	13	34	etc.
giving	$x = 2$	5	13	34	89	etc.
	$y = 1$	2	5	13	34	etc.

Obviously, the only acceptable solution is $x = 34$, $y = 13$.

20. AT THE ART STORE

The prices being given, we have:

$$210x + 330y + 462z + 770u + 1155v = 10001,$$

$$x, y, z, u, v \text{ all being integers} > 1$$

Dividing through by each of 11, 7, 5, 3, and 2:

$$\frac{x-2}{11}, \quad \frac{y-5}{7}, \quad \frac{2z-1}{5}, \quad \frac{2u-2}{3}, \quad \frac{v-1}{2} \text{ must be integers,}$$

then $$\frac{6z-3}{5} \text{ is an integer,}$$

so $$\frac{z-3}{5} \text{ and } \frac{u-1}{3} \text{ are integers.}$$

Hence, we can say:

$$x = 11a + 2, \quad y = 7b + 5, \quad z = 5c + 3, \quad u = 3d + 4, \quad v = 2e + 3,$$

where a, b, c, d, e must each be a positive integer or zero, thereby complying with condition that $x, y, z, u, v > 1$. Substituting these values, the original equation becomes:

$$2310(a + b + c + d + e) = 0$$

hence $$a = b = c = d = e = 0$$

so $$x = 2, \quad y = 5, \quad z = 3, \quad u = 4, \quad v = 3$$

The purchase, therefore, was:

2 frames	@	$ 2.10
5	@	$ 3.30
3	@	$ 4.62
4	@	$ 7.70
3	@	$11.55

22. DO IT YOURSELF

Say he had y tiles, each x inches square, area yx^2. $(y + 252)$ tiles, each $(x - \frac{3}{4})$ inches square, would have

area
$$\frac{(y + 252)(4x - 3)^2}{16} = yx^2,$$

whence
$$y = 168x - 189 + \frac{189}{8x - 3}$$

Now x and y are integers, so 189 is a multiple of $(8x - 3)$,
$$189 = 3^3 \cdot 7, \text{ so } (8x - 3) = 7, 9, 21, 27, 63, \text{ or } 189.$$
But, of these, integral x entails $(8x - 3) = 21$ or 189.

	$8x - 3 =$	21	189
with	$x =$	3	24
and correspondingly	$y =$	324	3844

Obviously, the table could not require 3844 24-inch tiles, so there were 324 tiles each 3 inches square.

23. WHO PLAYS SWITCH?

Say there were x players, and they played y games.

Each game, prior to the last, kitty gains $5(x - 6)\cancel{c}$, so, at start of last game, the kitty is $5(x - 6)(y - 1)\cancel{c}$.

After the final game, the kitty receives $5(x - 1)\,\cancel{c}$, and becomes $5\{(x - 6)(y - 1) + (x - 1)\} = 5(xy - 6y + 5)\cancel{c}$.

Jim, on the previous $(y - 1)$ games had lost $5(y - 1)\cancel{c}$. After clearing the kitty he ends up \$1 to the good.

So, $5(xy - 6y + 5) - 5(y - 1) = 100,$

whence $y(x - 7) =$ 14

Each player had won at least once, so $y \geq x$, and x and y are integers. Hence $y = 14$, $x - 7 = 1$, $x = 8$.

There were 8 players, and they played 14 games.

24. A TALE OF SOME CAKES

We derive the equation $12x + 14y + 17z = 200$, x,y,z being integers.

This has the general solution: $x = 11 - 4y + 17k$
$$y = y$$
$$z = 4 + 2y - 12k$$
k being any integer.

From x, we have $4y < 17k + 11$. From z, $4y > 24k - 8$, hence $k < 3$. Also, obviously $k \geq 0$.

Tabulate the resulting 5 sets of acceptable values:

$x =$	7	3	8	4	1
$y =$	1	2	5	6	11
$z =$	6	8	2	4	2
Total =	14	13	15	14	14

The mother knew the total, but was still in doubt. So the total was 14.

If, in reply to her question, Garry had said "Yes" then he would have bought either 7, 1, 6 OR 1, 11, 2. The details would have remained in doubt.

So he must have said "No" (in effect) and from that his mother knew that he had bought:

4 at 12¢, 6 at 14¢, and 4 at 17¢.

25. THE ROAD WENT THROUGH

Say the totals from each province, including foremen, were:

$$
\begin{array}{ll}
\text{Manitoba} & 17x + a \\
\text{Quebec} & 17y + b \\
\text{Ontario} & 17z + c
\end{array}
\left.\right\}
\begin{array}{l}
a,\, b,\, c \text{ being} \\
\text{integers,} \\
\text{all} < 17.
\end{array}
$$

Then, $17(x + y + z) + a + b + c = 600$

The "remainder," $(a + b + c)$ are divided into 2 groups:

    (1) a Manitoba men
    (2) b Ontario men, and 1 Quebec man.

These two groups are equal, and $b = 1$, so $c = a - 1$, whence $a + b + c = 2a$.

Hence, $17(x + y + z) + 2a = 600$,

with general integral solution: $\left.\begin{array}{l} x + y + z = 2k \\ a = 300 - 17k \end{array}\right\}$

But, $a >$ zero, and $a < 17$, so $k = 17$, hence $a = 11$.

Then, $x + y + z = 34$.

Now, as regards the foremen (included in the foregoing), there were $(x + 1)$ from Manitoba, $(y + 1)$ from Quebec, z from Ontario. These being all equal, $x + 1 = y + 1 = z$.

Hence, $3x + 1 = 34$, so $x = 11$, with $y = 11$, $z = 12$. This provides the required solution:

Manitoba	198 (including 12 foremen)
Quebec	188 (including 12 foremen)
Ontario	214 (including 12 foremen)

26. A BREAK FROM READING

Say he was reading page n.

There were $(n - 13)$ pages before his page, with sum:

$$\frac{(n - 13)(n + 12)}{2}$$

If the last page was page m, he had $(m - n)$ pages after his page, with sum:

$$\frac{(m - n)(m + n + 1)}{2}$$

These being equal, $(n - 13)(n + 12) = (m - n)(m + n + 1)$,

whence $(2m + 1)^2 - 2(2n)^2 = -623$

Now, $623 = 7 \cdot 89$, both factors being prime.

Hence (see Chapter 6) we can expect to find 2 distinct families of solutions for this Pellian equation.

We find these based on: $\left.\begin{array}{l} 2m + 1 = \ \ 5 \\ 2n = 18 \end{array}\right\}$ and $\left.\begin{array}{l} 2m + 1 = 23 \\ 2n = 24 \end{array}\right\}$

giving the general integral solutions:

$$\left.\begin{array}{l} 2m + 1 = \pm(5a \pm 36b) \\ 2n = \pm(18a \pm 5b) \end{array}\right\} \quad \text{and} \quad \left.\begin{array}{l} 2m + 1 = \pm(23a \pm 48b) \\ 2n = \pm(24a \pm 23b) \end{array}\right\}$$

where $a^2 - 2b^2 = 1$.

Tabulating the successive pairs of values we derive:

$m =$	13	28	43	82	92	173	258	483	541	etc.
$n =$	13	22	32	59	66	123	183	342	383	etc.
$m - n =$	0	6	11	23	26	50	75	141	158	etc.

He still had "nearly 150 pages" to read, so $m = 483$, $n = 342$ gives the required solution.

27. HIS PRIVATE ARMY

Say there were x^2 soldiers in the smaller of the two squares, and there would be y^2 in each of the five that Peter planned. x and y are integers.

Then,
$$x^2 + (x + 3)^2 = 5y^2,$$
whence
$$(2x + 3)^2 - 10y^2 = -9.$$

This Pellian equation (see Chapter 6) has two distinct families of integral solutions, based on:

$$(2x + 3) = 1, \quad y = 1, \quad \text{and} \quad (2x + 3) = 9, \quad y = 3.$$

Tabulating successive pairs of solutions, we have:

$$
\begin{aligned}
x &= 3 \quad\quad 19 \quad\quad 38 \quad\quad 174 \quad\quad \text{etc.} \\
y &= 3 \quad\quad 13 \quad\quad 25 \quad\quad 111 \quad\quad \text{etc.}
\end{aligned}
$$

Altogether, there were $5y^2$ soldiers, this total being considerably more than 1000. Also, the soldiers were arrayed on a large table, so $y = 111$ could not be an acceptable solution.

Hence, $x = 38$, $y = 25$, and there were 3125 soldiers.

28. THE GREAT BALL OF KAAL

Say diameters were: $x \geq y \geq (38 - x - y)$,
x, y being integers.

Then,
$$x^3 + y^3 + (38 - x - y)^3 = 8000$$
whence
$$(x + y)\{xy + 1444 - 38(x + y)\} = 15624,$$
$$15624 = 2^3 \cdot 3^2 \cdot 7 \cdot 31, \quad \text{and} \quad (x + y) \text{ is a factor of } 15624.$$

Because $x \geq y \geq (38 - x - y)$, $x + y \geq 26$,

also $x + y < 38$,   so $x + y = 28$ or 31 or 36.

With $x + y = 36$,   we have $xy = 358$,   with no solution.

With $x + y = 28$,   we have $xy = 178$,   with no solution.

With $x + y = 31$,   we have $xy = 238$,   hence $(x,y) = (17,14)$.

So the required diameters were 17, 14, and 7 kebals.

29. AFTER THE RUSH

The price for 1 goose and 1 duck must total an integral number of dollars, although there is no such limitation on their respective prices. Hence, we say:

$2x$ turkeys @ 700¢, x geese @ $(100y - z)$¢, y ducks @ z¢,

where $3x + y = 20$,   and x, y, z are integers.

Thence, $75x^2 - (850 - z)x - 5z + 2500 = 0$,

so $150x = 850 - z \pm \sqrt{(z^2 - 200z - 27500)}$.

Now, x and z are integers, so say $z^2 - 200z - 27500 = k^2$, k being an integer. Then,

$$(z - 100)^2 - k^2 = 37500 = 2^2 \cdot 3 \cdot 5^5$$

Also, $y \leq 17$,   so $z < 1700$, and $z - 100 < 1600$

Observing this last condition, now tabulate as follows:

$z - 100 + k =$	3750	1250	750	250
$z - 100 - k =$	10	30	50	150
$z - 100 =$	1880	640	400	200
$z =$	—	740	500	300
$k =$	—	610	350	50
$150x =$	—	720	700	500 or 600
$x =$	—	—	—	4
with $y =$	—	—	—	8

So Jake bought 4 geese @ $5, 8 turkeys @ $7, 8 ducks @ $3.

30. WHOSE HAT?

This can be solved diagrammatically, but the neat Boolean solution is more interesting.

We code the four members, in relation to the taking and losing of a hat, as:

	Taking	Losing
Andy	A	a
Bill	B	b
Charlie	C	c
Don	D	d

Obviously, $Aa = 0$, $Bb = 0$, $Cc = 0$, $Dd = 0$. Also, as Andy and Bill did not take each other's hats, $Ab = 0$, $Ba = 0$.

From the data given, we have three equations:

$$Ac \cdot Bd + Ad \cdot Bc = 1 \dots\dots\dots\dots (1)$$
$$Ca + Cb + Cd = 1 \dots\dots\dots\dots (2)$$
$$Da + Db + Dc = 1 \dots\dots\dots\dots (3)$$

Multiply (2) by (3) and strike out zero terms, i.e., $Da \cdot Ca$, $Db \cdot Cb$, and $Dc \cdot Cd$. We are left with:

$$Da \cdot Cb + Da \cdot Cd + Db \cdot Ca + Db \cdot Cd + Dc \cdot Ca + Dc \cdot Cb = 1 \dots (4)$$

Multiply (4) by (1), striking out any term that contains a, b, c, or d more than once. Also strike out any term that contains elements of the form $Cx \cdot Xc$ (re man whose hat Charlie took).

We are left with $Da \cdot Cb \cdot Ac \cdot Bd + Db \cdot Ca \cdot Ad \cdot Bc = 1$

So one of these terms represents the truth, and we can re-arrange each term to give: $Da \cdot Ac \cdot Cb \cdot Bd + Db \cdot Bc \cdot Ca \cdot Ad = 1$

Now, Don took the hat of the man who took the hat of the man who took Andy's hat. $Da \cdot Cb \cdot Ac \cdot Bd$ does not agree with this, therefore this term must have zero value.

Hence, we are left with $Db \cdot Bc \cdot Ca \cdot Ad = 1$.

So, Don took Bill's hat, Bill took Charlie's, Charlie took Andy's, and Andy took Don's hat.

32. THE DOLLAR DINNER

Say the dollar dinner had been selling through x weeks, and there were y girls.

$$\sum_1^z (x^3) = x^2(x + 1)^2/4 = y^4,$$

so $$x(x + 1) = 2y^2$$

whence $$(2x + 1)^2 - 2(2y)^2 = 1.$$

This is the simplest Pell equation (see Chapter 6), with solutions:

$2x + 1 = 3$	17	99	etc.
$2y = 2$	12	70	etc.
$x = 1$	8	49	etc.
$y = 1$	6	35	etc.

giving

But, $x < 26$, so the required solution is $x = 8$, $y = 6$, hence the total of those dinners sold was 1296.

33. THE STRUCTURE STILL STANDS

Say there were x^3 blocks in the cube part, $(y^2 - x^2)$ in the platform part, x and y being integers.

Then $y^2 - x^2 = 2x^3$, whence $x^2(2x + 1) = y^2$.

So, say $2x + 1 = z^2$, whence $2x = z^2 - 1$.

Here z must be odd, so say $z = 2k - 1$, whence $x = 2k(k - 1)$, $y = 2k(k - 1)(2k - 1)$.

From Ken's comment about the height, $k = 4$, making $x = 24$, and $x^3 = 13824$.

So in the complete structure there were 41472 blocks.

34. THE DOG LEG ROUTE

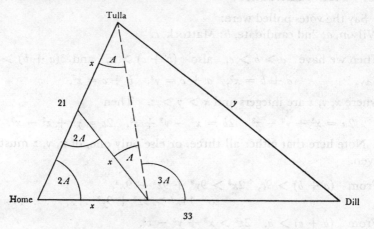

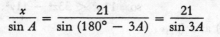

$$\frac{x}{\sin A} = \frac{21}{\sin (180° - 3A)} = \frac{21}{\sin 3A}$$

but, $\sin 3A = \sin A(2 \cos 2A + 1)$,   so $\cos 2A = \dfrac{21 - x}{x}$.

$$y^2 = 21^2 + 33^2 - 2 \cdot 21 \cdot 33 \cos 2A = 2223 - (3^3 \cdot 7^2 \cdot 11)/x.$$

x and y are both integers, so x is a factor of $3^3 \cdot 7^2 \cdot 11$. Also, $x > 21/3$, so $x > 7$ and obviously $x < 21$. Hence, $x = 9$ or 11.

Only with $x = 11$ do we have an integral value for y. So

$$x = 11, \quad \text{with } y = 30.$$

The dog leg route, then, was 33 miles.

35. THE ELECTION

Say the votes polled were:
Wilson, a; 2nd candidate, b; Mattock, c.

Then we have $a > b > c$,   also $(b + c) > a$,   and $(a + b) > 9c$.

Say, $a + b = x^3$, $a + c = y^3$, $b + c = z^3$,

where x, y, z are integers and $x > y > z$.   Then

$$2a = x^3 + y^3 - z^3, \quad 2b = x^3 - y^3 + z^3, \quad 2c = y^3 + z^3 - x^3.$$

Note here that either all three, or else only one, of x, y, z must be even.

From $(a + b) > 9c$, $2x^3 > 9y^3 + 9z^3 - 9x^3$,

$$\text{so} \quad 11x^3 > 9z^3 + 9y^3 \dots\dots\dots\dots\dots \quad (1)$$

From $(b + c) > a$, $2z^3 > x^3 + y^3 - z^3$,

$$\text{so} \quad 11x^3 < 33z^3 - 11y^3 \dots\dots\dots\dots \quad (2)$$

Combining (1) and (2),

$$33z^3 - 11y^3 > 9z^3 + 9y^3, \quad \text{so } 6z^3 > 5y^3$$

Similarly,

$$11x^3 - 9y^3 > 9z^3, \quad \text{and} \quad 9z^3 > 3x^3 + 3y^3, \quad \text{so } 2x^3 > 3y^3.$$

Also, as z is at least 1 less than y, $6z^3 > 5(z + 1)^3$,

so $z > 15$,   and $y > 16$.

Now tabulate values for y, with corresponding acceptable values
(if any) for z and x within the limitations $6y^3 > 6z^3 > 5y^3$, and
$6z^3 - 2y^3 > 2x^3 > 3y^3$.

For $y = 17$, 18, and 19 it will be found that correspondingly $z = 16$,
17, and 18, but without any integral values for x within the stipulated limits.   With $y = 20$, however, we have $6y^3 = 48000$ and
$5y^3 = 40000$, so $6z^3 = 41154$ and $z = 19$: also, $6z^3 - 2y^3 = 25154$ and
$3y^3 = 24000$, so $2x^3 = 24334$ and $x = 23$.

With $x = 23$, $y = 20$, $z = 19$, only one of these being even, the
relevant condition is satisfied.   Thence $a = 6654$, $b = 5513$, $c =
1346$, which provides the required solution.

36. WHAT'S A NUMBER?

We have $x^2 + 15 = y^2$, and $x^2 - 15 = z^2$, x,y,z being rational fractions with integral numerators and denominators.

Then, $30 = y^2 - z^2 = (y + z)(y - z)$, so y and z are both even or both odd.

Hence, $(y - z)$ must be even.   Say $y - z = 2k$

whence $y + z = \dfrac{15}{k}$

Then, $y = \dfrac{15}{2k} + k$,   and $z = \dfrac{15}{2k} - k.$

Substituting in $2x^2 = y^2 + z^2$, we get:

$$x^2 = \left(\frac{15}{2k}\right)^2 + k^2$$

This, in effect, is the familiar Pythagorean equation (see Chapter 6) with solution:

$$x = (a^2 + b^2)t, \quad \frac{15}{2k} = (a^2 - b^2)t, \quad k = 2abt.$$

Then, $t = k/2ab$, and substituting this we have:

$$x = \frac{(a^2 + b^2)k}{2ab}, \quad y = \frac{(a^2 + 2ab - b^2)k}{2ab}, \quad z = \frac{(a^2 - 2ab - b^2)k}{2ab}$$

where $\dfrac{(a^2 - b^2)k^2}{ab} = 15$

By quick trial it is seen that the simplest solution to this last equation is given by $a = 4$, $b = 1$, $k = 2$, whence

$$x = \tfrac{17}{4}, \quad y = \tfrac{23}{4}, \quad z = \tfrac{7}{4}.$$

So, the required fraction is $\tfrac{17}{4}$.

37. DUST UNTO DUST

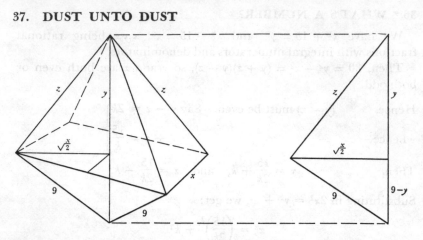

Referring to the diagrams, $9^2 - (9 - y)^2 = z^2 - y^2 = x^2/2$, x,y,z being integers.

Then $z^2 = 18y$, whence $y = 2k^2$, $z = 6k$, k being an integer.

Substituting for y and z, $\qquad x^2 = 8k^2(9 - k^2)$,

so $\qquad\qquad\qquad\qquad 18 - 2k^2 = t^2$, $\quad t$ being an integer,

whence $\qquad\qquad\qquad\qquad t^2 + 2k^2 = 18$,

which has only one integral solution: $t = 4$, $k = 1$, leading to $x = 8$, $y = 2$, $z = 6$.

Hence the dimensions were: base edge 8 feet, slant edge 6 feet, vertical height 2 feet.

38. WHERE THE GRASS GREW GREEN

Say a flower bed has sides, in feet:

$$(x + y), \quad (y + z), \quad (x + z), \quad x, y, z \text{ being integers.}$$

For any triangle, with sides A, B, C, the area is:

$$\sqrt{\{s(s - A)(s - B)(s - C)\}}, \quad \text{where} \quad 2s = A + B + C$$

So, for a flower bed, where we have $s = x + y + z$:

$$\text{area} = \sqrt{\{xyz(x + y + z)\}}, \quad \text{perimeter} = 2(x + y + z).$$

Hence, $xyz = 4(x + y + z)$, and we can say $x \geq y \geq z$.

Then, $\dfrac{xyz}{x + y + z} = 4$, and for $\dfrac{xyz}{x + y + z}$ to be a minimum z must be a minimum.

If $z = 1$, then $4(x + y + 1) = xy$, so $(x - 4)(y - 4) = 20$,

$$\text{giving} \quad x = 24 \quad\quad 14 \quad\quad 9$$
$$y = 5 \quad\quad\quad 6 \quad\quad\quad 8$$

and 3 triangles with sides: 29,   25,   6
                            20,   15,   7
                            17,   10,   9

If $z = 2$, then $4(x + y + 2) = 2xy$,

leading to 2 triangles with sides: 13,   12,   5
                                   10,   8,   6

If $z = 3$, then $4(x + y + 3) = 3xy$,

leading to 1 triangle: 13,   12,   5 (already noted)

Now, $\dfrac{4 \times 4 \times 4}{1 + 1 + 1} > 4$, so if $z > 3$ then $\dfrac{xyz}{x + y + z} > 4$.

So we have found all possible integral solutions, in the 5 sets of values:

6, 8, 10;   5, 12, 13;   9, 10, 17;   7, 15, 20;   6, 25, 29.

39. THE EXCURSION

If the single fare for a child was $y¢$, the complete fare structure must have been one of the following:

	Adult	Child	Adult	Child	Adult	Child	Adult	Child
Return	$4y$	$2y$	$4y-1$	$2y$	$4y-2$	$2y-1$	$4y-3$	$2y-1$
Single	$2y$	y	$2y$	y	$2y-1$	y	$2y-1$	y

Say there were $2x$ men, $2x$ children, 1 woman, and we can certainly assume $x < 25$. Then we consider the possible application of each fare structure.

(1) $y(9x + 4) = 5001$. But, $9x + 4$ cannot be a multiple of 3, so say $y = 3k$. Then $k(9x + 4) = 1667$, which is a prime. So there is no acceptable solution here.

(2) $x(9y - 1) + 4y - 1 = 5001$, hence $(9x + 4)(9y - 1) = 45014$. Now $45014 = 2 \cdot 22507$. To find factors of 22507, if any, would be laborious. But, if $9x + 4$ is a factor of 22507 then x must be odd, so we tabulate for odd values of x.

$x =$	1	3	5	7	9	11	13	15	17	19	21	23
$9x + 4 =$	13	31	49	67	8̶5̶	103	1̶2̶1̶	139	157	1̶7̶5̶	193	211

Before testing by actual division, strike out values that are obviously not factors of 22507, as shown. Then test remaining values by division. No acceptable solution is found here.

(3) $x(9y - 4) + 4y - 2 = 5001$, hence $(9x + 4)(9y - 4) = 45011$. Following the same procedure as in the previous case, we find that 139 is a factor of 45011, whence $x = 11, y = 49$.

(4) $x(9y - 5) + 4y - 3 = 5001$, hence $(9x + 4)(9y - 5) = 45016$. Now, 45016 can be factorized easily as $8 \cdot 17 \cdot 331$, leading to no acceptable solution.

So the required solution lies in the application (3), the details being:

11 men	@ $1.94 return......	$21.34
11 men	@ $0.97 single	$10.67
11 children	@ $0.97 return......	$10.67
11 children	@ $0.49 single	$ 5.39
1 woman	@ $1.94 return......	$ 1.94

40. SHARING THE PROFITS

Say, n partners with net profits x cents.

John's share left.............................. $\dfrac{(n-1)x}{n} + 1000$

Bruce's share left.............. $\dfrac{x(n-1)^2 + 1000n(n-1)}{n^2} + 1000$

nth share left balance of:

$$\frac{x(n-1)^n}{n^n} + 1000 \left[\left(\frac{n-1}{n}\right)^{n-1} + \left(\frac{n-1}{n}\right)^{n-2} + \ldots + 1 \right]$$

Say, balance after nth share was R, and $(n-1)/n = m$

Then $\quad R = xm^n + 1000(m^{n-1} + m^{n-2} + \ldots + m + 1)$
$\qquad\qquad = xm^n + 1000(1 - m^n)/(1 - m)$, where $m = (n-1)/n$

so $\qquad R = (x - 1000n)(n-1)^n/n^n + 1000n$

hence $\quad (x - 1000n) = (R - 1000n)n^n/(n-1)^n$, an integer.

Now, $n > 2$, so $n/(n-1)$ cannot be an integer, and there can be no common factor other than unity for n and $(n-1)$, hence $(R - 1000n)$ must be a multiple of $(n-1)^n$.

Say $\qquad R = k(n-1)^n + 1000n$, k being some integer

then $\qquad x = kn^n + 1000n$

Amount shared by partners was $(x - R)$ cents, i.e., 543607 cents

so $\qquad\qquad k[n^n - (n-1)^n] = 543607$

k is an integer, so $\qquad n^n - (n-1)^n \leq 543607$

and $\qquad n^n - (n-1)^n$ must be a factor of 543607.

Quick trial of $n = 3, 4, 5, 6$ provides no such factor. But $7^7 - 6^7 = 543607$, so we have $n = 7$ and $k = 1$

hence, $\qquad\qquad x = 823543 + 7000 = 830543$.

There were 7 partners, the net profits being \$8305.43. $R = 6^7 + 7000 = 286936 = 2^3 \cdot 13 \cdot 31 \cdot 89$, and each employee got "nearly \$100." So each employee's equal share was $2^3 \cdot 13 \cdot 89 = 9256$ cents = \$92.56, and there were 31 employees.

Appendix

A FUNDAMENTAL PROPERTY OF THE FIBONACCI SERIES

Theorem: Say the successive terms in the Fibonacci series be f_1, f_2, f_3, ... etc. Then the product of the $(n-1)$th and $(n+1)$th terms is alternatively k more and k less than the square of the nth term, where k is a constant depending only on the values of f_1 and f_2.

Proof: Say the difference, k, is not necessarily a constant.

Then, $\quad\quad\quad\quad f_{n-1} \cdot f_{n+1} = f_n^2 + a$, a being some number.

By definition, $\quad\quad\quad f_{n+1} = f_n + f_{n-1}$, for $n > 1$,

so $\quad\quad\quad f_{n-1}(f_n + f_{n-1}) = f_n^2 + a$

i.e. $\quad\quad\quad f_n^2 - f_{n-1}^2 = f_n \cdot f_{n-1} - a$, and similarly

$\quad\quad\quad\quad f_{n-1}^2 - f_{n-2}^2 = f_{n-1} \cdot f_{n-2} - b$, say.

Adding, $\quad\quad f_n^2 - f_{n-2}^2 = f_{n-1}(f_n + f_{n-2}) - (a + b)$.

Dividing through by $(f_n + f_{n-2})$, we get

$$f_n - f_{n-2} = f_{n-1} - \frac{a+b}{f_n + f_{n-2}}$$

but, $\quad\quad\quad f_n - f_{n-2} = f_{n-1}$ by definition,

hence $\quad\quad\quad a + b = 0$, so $a = -b$.

This is independent of n (for $n > 1$) so must be true for all values of n, hence the absolute value of a is constant.

We now have $\quad f_{n-1} \cdot f_{n+1} - f_n^2 = \pm k$ (a constant)

then $\quad\quad\quad\quad\quad f_3 \cdot f_1 - f_2^2 = (f_2 + f_1)f_1 - f_2^2$,

so $\quad\quad\quad\quad\quad k = \pm(f_2^2 - f_1^2 - f_1 \cdot f_2)$,

which is dependent only on values of f_1 and f_2, thus completing the proof of the theorem.

Note: In the familiar Fibonacci series, $f_1 = f_2 = 1$, whence $k = \pm 1$.

INDEX